SpringerBriefs in Computer Science

T0213682

More information about this series at http://www.springer.com/series/10028

Tingting Yang • Xuemin (Sherman) Shen

Maritime Wideband Communication Networks

Video Transmission Scheduling

 Springer

Tingting Yang
Navigation College
Dalian Maritime University
Dalian
China

Xuemin (Sherman) Shen
Department of Electronic and Computer Engineering
University of Waterloo
Waterloo, Ontario
Canada

ISSN 2191-5768   ISSN 2191-5776 (electronic)
ISBN 978-3-319-07361-3   ISBN 978-3-319-07362-0 (eBook)
DOI 10.1007/978-3-319-07362-0
Springer Cham Heidelberg New York Dordrecht London

Library of Congress Control Number: 2014940143

Printed on acid-free paper

Springer is part of Springer Science + Business Media (www.springer.com)

Preface

There has been increasing interest in the domain of emerging maritime wideband communication networks, which could be a low-cost alternative for current maritime satellite system. It is envisioned that building up a "maritime highway" system will greatly contribute to the maritime distress, urgency, safety, and general communications, and thereby facilitate a myriad of attractive applications related to monitoring (e.g., large capacity data such as surveillance videos being uploaded after collecting from bridge, engine room or other critical regions of a vessel), safety (e.g., maritime safety information dissemination), infotainment (e.g., mobile office and multimedia data download and upload, especially for cruise industry), and cargo online management (e.g., real-time cargo status notification and handling management). With rapidly evolving applications, maritime wideband communication networks can not only revolutionize the navigation pattern to be safer and more efficient, but also extend wideband services from the land to the ocean with lower expense.

This brief aims at providing valuable insights on the video transmission scheduling for the maritime wideband networks. Notably, the alternative eco-friendly and renewable green energy in maritime environment will make the scheduling issue more challenging and fascinating. To the best of our knowledge, this is the first work in the form of monograph to do such investigation facing distinguished challenges with unique characteristics imposed in maritime wideband networks. The problem is of great importance since related fundamental guidance is very limited. In Chap. 1, we give a brief introduction to maritime wideband communication networks, maritime video applications and challenges to video transmission scheduling, based on identifying unique characteristics of this dedicated system. A comprehensive survey is then provided pertaining to maritime wideband networks, video transmission scheduling and energy modeling based video transmission scheduling in Chap. 2. In Chap. 3, video transmission scheduling is studied aiming at maximizing the weights of uploaded video packets. A store-carry-and-forward routing mechanism is designed to address the intermittent connectivity of maritime wideband communication networks. In Chap. 4, an energy and content aware scheduling scheme is proposed to maximize the delivered video packets throughput in green-energy-powered maritime wideband networks, based on the analysis of green energy buffer model. Especially, the simulations in this brief present real cases studies, which all depend on real ship route traces obtained from navigation software BLM-Shipping. Finally, we draw conclusions and highlight future research directions in Chap. 5. This brief is meant as a petri dish to provide valuable guidance and new ideas in the field of video transmission scheduling for future maritime wideband networks.

The authors would like to thank Hao Liang, Nan Cheng, Zhongming Zheng, Ruilong Deng, Ning Lu, Ning Zhang, Xiaoxia Zhang and other Broadband Communications Research Group (BBCR) members at the University of Waterloo, for their contributions in the presented research works. They would also like to thank the editors at Springer US: Susan Lagerstrom-Fife and Jennifer Malat for their help throughout the publication preparation process. This work was supported by China Postdoctoral Science Foundation under Grants 2013M530900; International Postdoctoral Academic Exchange Fellowship Program, and NSERC, Canada.

Contents

Acronyms

AIS	Automatic Identification System
APs	Access Points
CCTV	Vessel Closed-circuit Television
DTNs	Delay Tolerant Networks
EVTMP	Energy and Content Aware Vessel Throughput Maximize Problem
FBB	Fleet Broadband
FIFO	First-Input-First-Output
IGTJRS	Interval Graph Theory based Job Relay Selection
IMO	International Maritime Organization
LTE	Long-Term Evolution
MAC	Media Access Control
MW2IS	Maximum Weight 2-Independent Set
QoS	Quality of Service
TDMA	Time Division Multiple Access
TDF	Throughput-Delay-Fairness
TMTP	Time-capacity Mapping based Two Phase
TMP	Throughput Maximization Problem
VANETs	Vehicular Ad hoc Networks
V2V	Vessel-to-Vessel
V2S	Vessel-to-Shore side
V2C	Vessel-to-Maritime Centre
VBR	Variable Bit rate
VHM	Vehicle Health Monitoring
VTMP	Vessel Throughput Maximization Problem
WLANs	Wireless Local Area Networks
WSN	Wireless Sensor Network

Chapter 1
Introduction

Maritime wideband communication networks target the incorporation of wireless communications and informatics technologies into the navigation transportation system, which revolutionize the navigation pattern to be safer and more efficient. Meanwhile, with the environmental consciousness and demand of efficient energy consumption, the eco-friendly and renewable green energy is envisioned as a promising alternative energy source in maritime communication networks. In this chapter, we first introduce the motivation and envision of maritime wideband communication networks, as well as the potential applications and challenges showcasing of this dedicated network, based on the analysis of unique characteristics. Finally, we summarize our contributions related to video transmission scheduling in maritime wideband communication networks.

1.1 Overview of Maritime Wideband Networks

Emerging maritime wideband communication networks are envisaged to facilitate a myriad of attractive applications such as maritime distress, urgency, safety, and general communications [1]. There are two major impetuses pushing forward the research of maritime wideband networks. The first one is the urgent need to provide wideband network, and improve efficiency and safety of maritime transportation systems catering to the ever-increasing mobile data demand at sea. In particular, large capacity data such as surveillance videos collected from bridge, engine room or other critical regions of a vessel, can be efficiently delivered via such a system, which is crucial to maritime administrative authority on land. Moreover, as an essential part of our daily life, Internet access on-board is expected anytime and anywhere.

Furthermore, safety related information and multimedia data could also be disseminated via this network. In other words, it extends wideband services from the land to the ocean. However, the state-of-the-art maritime communication system, christened as Global Maritime Distress and Safety System, comprises of terrestrial and satellite systems [2]. Only the advanced Fleet Broadband (FBB) system which belongs to satellite system, can be enabled to establish wideband transmission with data rate up to 432 kbps. Nonetheless, the high capital expenditure to

T. Yang, X. (Sherman) Shen, *Maritime Wideband Communication Networks,*
SpringerBriefs in Computer Science,
DOI 10.1007/978-3-319-07362-0_1, © The Author(s) 2014

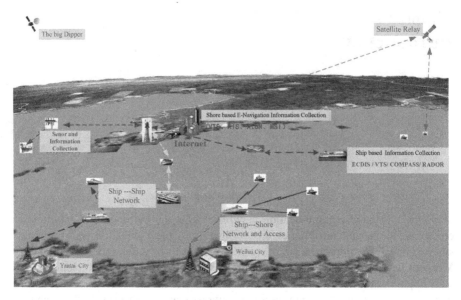

Fig. 1.1 An overview of maritime wideband networks

launch satellites results in high service cost by satellite system (e.g., voice service costs 13.75 U. S. dollars per minute). Consequently, the cost of conveying large capacity videos could be prohibitive. It is an imperious demand to develop a novel and cost-effective wide-band maritime communication network by innovative communication technologies. Relying on different communication technologies, the maritime wideband communication networks basically comprise three types of communications, Vessel-to-Vessel (V2V), Vessel-to-Shore side (V2S), and Vessel-to-Maritime Centre (V2C) through satellite, which are motivated by the definitions in Vehicular Ad hoc Networks (VANETs) [3], as shown in Fig. 1.1. It perfectly matches the emerging E-Navigation strategy initiated by International Maritime Organization (IMO), a led concept based on the harmonisation of marine navigation systems, could support shore services driven by user needs [4].

1.2   Maritime Video Applications

Leveraging high-rate Internet access for vessels, maritime wideband communication networks thereby could be envisioned not only to cater to the ever-increasing Internet data demand at sea, but also to facilitate a myriad of attractive applications. For example, safety-related monitoring and dissemination (e.g., large capacity data such as surveillance videos be uploaded after collecting from bridge, engine room or other critical regions of a vessel, as well as maritime safety information could be achieved dissemination via this system), cargo online management (e.g., real-time cargo status notification and handling management) and infotainment on the ocean (e.g., mobile office and multimedia data download and upload, especially for cruise industry), could be innovatively achieved via the dedicated maritime wideband communication networks. In this monograph, we focus on video applications. Based on the following three types of maritime video applications, the advanced maritime networks could not only make the navigation transportation system to be safer and more efficient, but also revolutionize the experience of the passengers with media-rich infotainment on the ocean.

Safety-Related Monitoring The threat of maritime piracy has mushroomed enormously in the past few years, especially for the sea areas mentioned following are piracy affected areas where the terror and threat of sea pirates has reached looming proportions: Malacca Straits, South China Sea, Gulf of Aden and so on. The news channels on a daily basis have several incidents to report about pirates attacking crew and looting the vessel or hijacking a ship, and even causing harm to the crew when their ransom demands are not met by the authorities [5]. Internet access can indubitably enrich safety-related applications. For example, surveillance videos collected from the interior and exterior of the vessels could be transmitted to the maritime administrative authority on land via the maritime wideband networks. Simultaneously, some safety related information (e.g., navigational warning and meteorological warning), multimedia data and command or documents broadcast by authority could also be disseminated via this system.

Real-Time Cargo Monitoring and Management In the networks of connected vessel or vessel-shore, real-time cargo information generated by the on-board sensors, control system and vessel-borne computer can be effectively disseminated to shore-based authority or among vessels in proximity. Therefore, maritime wideband communication networks combining with modern logistics provide the flexible flow of cargo management between the origin port to destination port. If the physical property obtained from the monitoring videos of atmosphere-sensitive cargo has changed obviously during the voyage, all these information could be collected, monitored and delivered to the growers, manufacturers, importers/exporters, ocean carriers, and retailers on shore timely. Hence, the cargo may be distributed at different ports under the decision on shore. The real-time cargo visibility and management could be achieved with competitive advantage, while, the need for transparency throughout the shipping industry will reach a completely new level.

Multimedia Data Uploading or Downloading With rapidly increasing demand of Internet access, the maritime wideband communication networks enable passengers to upload and download multimedia data on board, by V2S communication. Imagine that during a short tour in Aegean Sea located northeast of Mediterranean Sea, you are exploring the paradise. The multitudinous islands, spreading all over let you stay hooked and as if you were in the beautiful love legend. All the photos and videos you were taken and amazing scenery reminds you to share them with your beloved. Through the maritime wideband communication networks, it could be possible to connect you to the Internet achieving multimedia data uploading or downloading.

1.3   Challenges to Video Transmission Scheduling

Fascinated by visions of maritime wideband networks, the academia, industry and government institutions have initiated some activities recent years. An introduction of project of wireless-broadband-access for Seaport (WISEPORT) and TRITON in Singapore can be found in [6, 7]. Wireless broadband access could be achieved with rate up to 5 Mbps, based on Worldwide Interoperability for Microwave Access (WiMAX) technology [6]. Taking advantage of its high data rate and large coverage area, WiMAX technology has been approved to be a candidate to satisfy the increasing demand of wideband data traffic at sea [8]. Furthermore, the US Navy ships have used advanced Fourth Generation Long-Term Evolution (4G LTE) broadband service since 2011 [9]. Hence, the realization of maritime communications can resort to advanced technology, e.g., WiMAX technology or LTE technology, etc. However, the coverage range is still limited (e.g., with a range of approximately 20 nautical miles from coastline). Moreover, the wireless channels are occasionally deteriorated due to the obstacles in terms of sea clutters, and the period of wireless connection is short because of a limited number of infostations deployed shore-side. Consequently, a continuous end-to-end path may not be available in the maritime environment. One natural question is whether it is possible to improve transmission capability of maritime networks by employing any advanced techniques or sophisticated strategies. After significant progress that has been made to further the investigation on network designing, the answer is positive. An innovative complementary scheme is store-carry-and-forward packet delivery in delay tolerant networks (DTNs) [10], which can efficiently utilize node mobility statistics and permit other nodes to store, carry, and deliver data packets once a communication opportunity arises.

Green energy refers to eco-friendly and sustainable energy sources, e.g. solar, wind, hydro, and modern biomass, etc. The cost of maritime wireless networks at sea including deployment, energy and maintenance is high due to geographic constraints. With the emerging of green energy, it is possible to cost-effectively construct green-energy-powered maritime wireless networks. The advances of green wireless networks provide an alternative energy for maritime wireless networks, which can significantly reduce the cost of maritime wireless networks establishment and maintenance. However, unlike traditional energy provided by electrical grid, the green energy is replenished from nature and highly depends on environment. Thus, the fundamental design criterion in the network deployment and resource management is shifted from energy efficiency to energy sustainability due to the sustainable nature of green energy.

In this monograph, we design a broadband wireless network/store-carry-and-forward interworking maritime wideband communication system, which utilizes TDMA MAC protocol widely used in WiMAX and LTE technology, explicitly devised to overcome above limitations. In addition, video transmission scheduling counting the dynamic energy modeling is investigated. We consider the Vessel Closed-circuit Television (CCTV) Systems [11], and surveillance video clips are generated at regular durations, which are further divided into packets. Each packet has a release time, a deadline, and a weight. Weight is the quotient to value the importance of video clips or the significance to the administrative authority. In this brief, deadline means playback deadline, which is defined as a time point that the video packet will be decoded successfully by land-based authority if it is delivered before that moment. Meanwhile, the maritime wideband communications scenario utterly distinguishes from the vehicular networks [12, 13] and high-speed trains [14]. In vehicular networks, the traces of vehicles are nondeterministic and dynamically changing. In contrast, the traces of vessels are deterministic or predictable since ship routes are relatively stable and known *a priori*, due to the Ships' Routeing scheme recommended by IMO especially the passenger vessels and cargo vessels for the water area with large navigation density, as we do not consider vessels with lower tonnage such as fishing boats. In terms of a high-speed train, the train schedule is deterministic and the train trajectory is one-dimensional. Thus, the main challenges of video transmission scheduling in maritime scenario are summarized as follows:

- **Intermittent networks connection and limited *time window*.** With the goal of throughput maximization, how to schedule video packets to be relayed by other vessels needs to be investigated, suffering limited *time window* within which video packets could be transmitted to shore-side infostations. These issues are complicated and challenging as the links from vessel to infostations are highly dynamic and subject to periodic disconnections as a vessel sails en route.
- **One-dimensional or two-dimensional network topology.** The network topology in maritime communication networks could be either one-dimensional (without relay) or two-dimensional (with relay). How to schedule video packets in two-dimensional and intermittently connected maritime communication system with deterministic global knowledge is still an open issue.
- **Energy sustainability.** Green energy highly depends on its position, local weather and time, which makes the green energy inherently variable or even intermittent with time. Thus, the fundamental design criterion and the main performance metric under the scenario of green-energy-powered maritime wireless networks are shifted from energy efficiency to energy sustainability. The new dimension of energy sustainability makes the video transmission scheduling in green energy maritime wireless communication networks more challenging.

1.4 Our Contributions

Video transmission scheduling for maritime wideband networks is still embryonic. Due to the particular characteristics of maritime communication scenario, many research fruits referring to video transmission scheduling for land-based paradigms need to be re-visited. To the best of our knowledge, our work is the first to investigate such video transmission scheduling issues in maritime communication networks, even in maritime networks powered by renewable energy sources, with the goal of throughput maximization. Specifically, the contribution of this brief is three-fold:

- We develop a framework for vessel surveillance video uploading via a maritime wideband communication network. A broadband wireless network utilizing Time Division Multiple Access (TDMA) based media access control (MAC) protocol is employed to establish a shore-side network infrastructure. The realization of intermittent network connectivity in maritime communication network can resort to packet store-carry-and-forward routing mechanism.
- The video transmission scheduling problem is formulated to maximize the weights of uploaded video packets. Time-capacity mapping technique is used to transform the original scenario with intermittent network connectivity into a virtually continuous scenario. Two algorithms are proposed to offer important guidelines on video transmission scheduling.
- We formulate the energy and content aware vessel throughput maximization problem in green-energy-powered maritime wireless networks. The energy buffer of shore-side base station and DTN nodes are modeled as a $G/G/1$ queue, and a diffusion approximation method is engaged to investigate transient states. As a retreat when optimal algorithm is out of reach, two heuristic algorithms are proposed by scheduling the video packets delivered subject to the energy constraint, based on energy buffer modeling.

The remaining of this brief is organized as follows. In Chap. 2, we present the background information and literature survey related to maritime wideband networks, video transmission scheduling and energy modeling based video transmission scheduling. Chapter 3 investigates efficient scheduling for video transmission in maritime wideband communication networks. We jointly consider energy efficiency and content aware video transmission issues of sustainable maritime networks in Chap. 4. We conclude this brief and give a direction of maritime wideband communication networks for future research in Chap. 5.

References

1. T. Yang, H. Liang, N. Cheng, and X. Shen, "Towards video packets store-carry-and-forward scheduling in maritime wideband communication," in *Proc. IEEE GLOBECOM*, Dec. 2013, pp. 1–6.
2. I. Maglogiannis, S. Hadjiefthymiades, N. Panagiotarakis, and P. Hartigan, "Next generation maritime communication systems," *International Journal of Mobile Communications*, vol. 3, no. 3, pp. 231–248, March 2005.
3. N. Lu and X. Shen, *Capacity Analysis of Vehicular Communication Networks*. SpringerBrief, 2013.
4. E. Mitropoulos, "E-navigation: a global resource. seaways," *Journal of the Nautical Institute*, March 2007.
5. Sharda, "What is a group of ships called?" http://www.marineinsight.com/marine/marine-piracy-marine/10-maritime-piracy-affected-areas-around-the-world/, Aug. 2011.
6. Cellular-news, "Maritime WiMAX network launched in singapore," http://www.cellular-news.com/story/29749.php, March 2008.
7. J. S. PATHMASUNTHARAM, H. WANG, P.-Y. KONG, C.-W. ANG, Y. GE, S. WEN *et al.*, "Triton: high-speed maritime wireless mesh network," *IEEE Wireless Communications*, pp. 134–142, Oct. 2013.

8. V. Hoang, M. Ma, R. Miura, and M. Fujise, "A novel way for handover in maritime WiMAX mesh network," in *Proc. IEEE ITS*, 2007, pp. 1–4.
9. "Naval ships CCTV systems," http://www.km.kongsberg.com/ks/web/nokbg0240.nsf/AllWeb /9387546AF26FE067C125776C003DE5A9.
10. K. Fall, "A delay-tolerant network architecture for challenged internets," in *Proc. ACM Applications, technologies, architectures, and protocols for computer communications*. ACM, Aug. 2003, pp. 27–34.
11. K. John, "US Navy Ships To Get 4G LTE Broadband - Will Commercial Vessels Be Next?" http://gcaptain.com/navy-ships-4g-lte/, June 2012.
12. Q. Yan, M. Li, Z. Yang, W. Lou, and H. Zhai, "Throughput analysis of cooperative mobile content distribution in vehicular network using symbol level network coding," *IEEE Journal on Selected Areas in Communications*, vol. 30, no. 2, pp. 484–492, Feb. 2012.
13. M. J. Khabbaz, W. F. Fawaz, and C. M. Assi, "Modeling and delay analysis of intermittently connected roadside communication networks," *IEEE Transactions on Vehicular Technology*, vol. 61, no. 6, pp. 2698–2706, July 2012.
14. H. Liang and W. Zhuang, "Efficient on-demand data service delivery to high-speed trains in cellular/infostation integrated networks," *IEEE Journal on Selected Areas in Communications*, vol. 30, no. 4, pp. 780–791, May 2012.

Chapter 2
Background and Literature Survey

The study of new maritime wideband communication network commenced to be attracted attention recently. Although the research on video transmission scheduling in maritime wideband communication network is still in the early stage, a large number of counterparts on land-based network have shown up, which could lay down a solid foundation for maritime communication networks. Moreover, little literature with emphasis on video transmission scheduling problem in DTN maritime networks is presented. Many fundamental research issues have not been well studied. We categorize the existing works in the literature related to our works into three research issues.

2.1 Preliminaries: Milestone of Maritime Wideband Network

Emerging maritime wideband network has attracted significant attention in recent years. The line of project investigation began with TRITON [1], in which a wireless mesh network to support multi-hop data delivery in maritime network is investigated. However, this paper does not pay much attention about the scenario that vessels are in low density that DTN-based scheme should be utilized. Resorting cognitive radio technology to maritime scenario, a cognitive maritime mesh/ad hoc network is designed in [2]. While the WiMAX-based mesh technology for ship-to-ship communications with DTN features is explored in [3], and the performance between regular routing protocols and DTN routing protocols is compared. Based on a theoretical model to analyze the ships encounter probability distribution, the data delivery ratio from ships to the BS is derived [4]. In [5], the performance of file delivery in maritime DTN network is investigated. The transmission opportunities are existed only when there is direct link between the two vessels, while in our work, DTN throw-box is employed to transiently store data and raise the transmission opportunity. An approach of utilizing multi-hop WiMAX and mesh network is proposed in [6], to provide Internet access to the Mediterranean Sea without the help of satellite. The MAC and routing schemes that are suitable for such a scenario are investigated, with the network connectivity analyzed. However, it can be considered as a special case because the vessel density is high and Internet access is assumed to be available

T. Yang, X. (Sherman) Shen, *Maritime Wideband Communication Networks,*
SpringerBriefs in Computer Science,
DOI 10.1007/978-3-319-07362-0_2, © The Author(s) 2014

at anytime via the multi-hop mesh network. Besides some classic scheduling algorithms, we compare our work with the existing maritime DTN/none-DTN algorithms based on [1, 5] in Chap. 3.

2.2 Video Transmission Scheduling

Video transmission could provide safety-related monitoring for vessels as above mentioned. Scheduling focuses on decision-making, which plays a particularly important role to ensure that the new telecommunications network effectively meets the needs of the subscriber and operator. As real-time multimedia services have stringent quality-of-service (QoS) requirements to maintain user satisfaction, high quality video streaming over variable bit rate (VBR) channels represents several fundamental challenges in engineering [7]. For video transmission delivery, although there are very few research works in the literature for maritime scenario, a variety of research works on the land communication area appear in the literature. In [8], Fu *et al.* formulated the dynamic scheduling problem as a Markov decision process that explicitly considers the users heterogeneous multimedia data characteristics (e.g. delay deadlines, distortion impacts and dependencies etc.) and time-varying channel conditions. A directed acyclic graph is applied to express the transmission priorities between the different video units. In [9], Pahalawatta *et al.* presented a cross-layer packet scheduling scheme that streams pre-encoded video over wireless downlink packet access networks to multiple users, by designing the user utilities as a function of the distortion of the received video. A fair-scheduling algorithm for the transmission of video frames over wireless links is developed in [10], based on the occupancy of the video decoder buffer. In [11], an energy-efficient algorithm is presented for the video transmission scheduling in wireless P2P live streaming system, to minimize the playback freeze-ups among peers. Muhammad and Zhuang [12] proposed an energy and content-aware video transmission framework that incorporates the energy limitation of mobile terminals and the QoS requirements of video streaming applications. From aforementioned references, video transmission scheduling issues have gained an increasing popularity among various mobile applications.

Generally, video could be regarded as one type of large data, disregarding the unique characteristics. With respect to data transmission delivery, in [13], Yan *et al.* developed a theoretical model to compute the achievable throughput of cooperative mobile content distribution in vehicular ad hoc networks. The IEEE 802.11 p MAC protocol is proposed for video broadcasting in metro passenger communication system, which is specially designed for high-speed trains with a speed up to 360 km/h [14]. In [15], Liang *et al.* proposed a semi-Markov decision process based service model to efficiently manage inter-domain resource allocation in mobile cloud networks. In [16], Cheng *et al.* proposed a vehicle-assisted data delivery method for smart grid applications, based on optimal stopping theory. In [17], three alternatives of vehicular access infrastructure are designed, and the data transmission scheduling based on capacity-cost tradeoffs for different schemes are analyzed. In [18], Liang and Zhuang investigated on-demand data services for high-speed trains, based on Smith ratio and exponential capacity algorithms. In [19], the two-phase algorithm is leveraged in a joint time slot, power control and rate assignment problem in mobile wireless sensor network. Similar research efforts on transmission and access scheduling in cognitive radio networks can be found in [20, 21]. However, the non-flourish video (or even data) transmission scheduling works of maritime network, contrast sharply with counterpart on land communication network.

In Chap. 3, we utilize the job-machine scheduling method to address Throughput Maximization Problem (TMP) problem. A plenty of TMP algorithms remark machine independent issue. However, duo to limited communication periods, mobile users might contact with APs with machine-dependent time window. Unrelated machines scheduling problem has been studied in [22] firstly, where the time indices in terms of release time, deadline and processing time are all machine dependent. Meanwhile, the scheduling issue relating to machine-dependent time indices is investigated in [23]. However, the scenario is distinct from ours, where the coverage area of APs is consecutive and each user has only one job to transmit. The two-phase algorithm proposed in [19] is utilized to jointly allocate the resources of timeslot, power and rate in mobile WSN. However, it is a parallel machine issue, without considering cooperation transmission.

In this brief, we focus on designing ship-shore video transmission scheduling schemes in DTN maritime communication networks, with the goal of throughput maximization while considering cooperation between different vessels and the heterogeneity of the different videos.

2.3 Energy Modeling Based Video Transmission Scheduling

Driven by environment concerns, the next generation wireless networks are envisioned to make use of renewable energy, which can be harvested from natural and renewable resources, such as solar, wind, tides, etc. Motivated by the relatively high performance-cost ratio, solar and wind power are two of the most common energy sources that have been extensively used by wireless networks, especially by the network infrastructure. For instance, the Green Wifi initiative has developed a low cost, solar-powered, standardized Wifi solution for providing Internet access to developing areas [24]. The wind-powered wireless mesh networks are also applied for emergency network deployment after disasters [25]. However, most existing works concern about research issues under the scenario of traditional energy powered maritime wireless networks. Renewable energy sources are intrinsically dynamic with unstable availability and time varying capacity. As such, when renewable energy is used to power maritime communication networks, its dynamic and unreliable nature will affect the transmission efficiency. Accordingly, these impediments impose significant challenges on the maritime network resource management.

Designing an accurate analytic energy model is an effective method to enhance the energy efficiency. Thus, this energy model should be able to predict the remaining capability of powered devices, with a crucial issue of how to figure out the charging and discharging models. In [26], a more accurate energy model for wireless sensor node is proposed, and an optimal design method of energy efficient wireless sensor node is described, taking into account of the energy dissipation of circuits in practical physical layer. In [27], a novel energy model for batteries and study the effect of battery behavior on routing in wireless ad hoc networks is introduced, by proposing an online computable discrete-time mathematical model to capture battery discharging behavior. In [28], a detailed analytical model for estimating the total energy consumed to exchange a packet over a wireless link is provided, by considering details such as consumed energy by processing elements of transceivers, packet retransmission, reliability of links, size of data packets and acknowledgments, and also the data rate of wireless links. Similar research efforts on energy modeling can be found in [29, 30].

With respect to energy modeling based video (or data in broader sense) transmission scheduling in green wireless communication, although there are very few works related to maritime communication networks, land-based green wireless communication networks have sprung up around the world recently. Several works address video transmission scheduling over time-varying VBR channels, including the rate-distortion optimized video packet scheduling [31] and routing [32]. However, most of those approaches emphasize the network optimization in terms of throughput, delay and delay jitter, without considering the consequent video quality from the subscriber's perspective, which motivates our work. In [33], a mathematical framework is developed to study the impact of network dynamics on the perceived video quality. The close-form expressions of the video quality are given in terms of start-up delay, playback and packet loss, which provide better clue for our work in Chap. 4. Referring to data transmission scheduling, the authors of [34] identified that green-energy-powered APs provide a cost-effective solution for wireless local area networks (WLANs). In [35], the throw-box is assumed to be able to last for a certain period of time, which can calculate the average power from the capacity of the batteries or harvesting energy from solar panels. In [36], network deployment and resource management issues are investigated in the context of green mesh networks. A placement solution seeking paths with the minimum energy depletion probability is proposed to improve the network sustainability while ensures that the energy and QoS demands of mobile users can be fulfilled. In [37], a network planning problem in green wireless communication network is studied. The relay nodes placement and sub-carrier allocation (RNP-SA) issues are jointly formulated. Authors proposed top-down/bottom-up algorithms to minimize the number of APs powered by renewable energy sources with satisfying the quality of service (QoS) requirement of users.

An accurate analytical model of the power dynamics could effectively adapt to the routing decisions in an energy sustainable wireless network. Most existing works use residual energy of the mean energy charging rate for resource allocation, by either assuming that the energy charging rate is known *a priori* or using an oversimplified model. However, the charging in reality is known to be a complicated and dynamic process due to the restricted energy charging capabilities and diverse charging environments which is usually location dependent [38]. In Chap. 4, scheduling and routing schemes are developed to optimize the video transmission throughput and energy sustainability, based on dynamic $G/G/1$ energy queueing model.

2.4 Summary

This chapter has surveyed the existing literature pertaining to maritime wideband network and video transmission scheduling approaches. It has also presented a comprehensive overview of energy modeling based video transmission scheduling. Extensive comparisons of existing results have been done to reach a better understanding of the importance of our work.

References

1. J. Shankar Pathmasuntharam, P.-Y. Kong, M.-T. Zhou, Y. Ge, H. Wang, C.-W. Ang, W. Su, and H. Harada, "TRITON: High speed maritime mesh networks," in *Proc. IEEE PIMRC*, 2008, pp. 1–5.
2. M. T. Zhou, and H. Harada, "Cognitive maritime wireless mesh/ad hoc networks," *Journal of Network and Computer Applications*, vol. 35, no. 2, pp. 518–526, March 2012.
3. H.-M. Lin, Y. Ge, A.-C. Pang, and J. S. Pathmasuntharam, "Performance study on delay tolerant networks in maritime communication environments," in *Proc. IEEE OCEANS*, 2010, pp. 1–6.
4. S. Qin, G. Feng, W. Qin, Y. Ge, and J. S. Pathmasuntharam, "Performance modeling of data transmission in maritime delay-tolerant-networks," in *Proc. IEEE WCNC*, 2013, pp. 1109–1114.
5. P. Kolios, and L. Lambrinos, "Optimising file delivery in a maritime environment through inter-vessel connectivity predictions," in *Proc. IEEE WiMob*, 2012, pp. 777–783.
6. V. Friderikos, K. Papadaki, M. Dohler, A. Gkelias, and A. H., "Linked waters," *IEEE Communications Engineer*, vol. 3, no. 2, pp. 24–27, Apr. 2005.
7. B. Girod, J. Chakareski, M. Kalman, Y. Liang, E. Setton, and R. Zhang, "Advances in network-adaptive video streaming," *Wireless Commun. Mobile Comput.*, vol. 2, no. 6, pp. 549–552, June 2002.
8. F. Fu and M. van der Schaar, "Structural solutions to dynamic scheduling for multimedia transmission in unknown wireless environments," *arXiv preprint arXiv:1008.4406*, Aug. 2010.
9. P. Pahalawatta, R. Berry, T. Pappas, and A. Katsaggelos, "Content-aware resource allocation and packet scheduling for video transmission over wireless networks," *IEEE Journal on Selected Areas in Communications*, vol. 25, no. 4, pp. 749–759, May 2007.
10. M. Hassan, T. Landolsi, and M. Tarhuni, "A fair scheduling algorithm for video transmission over wireless packet networks," in *Proc. IEEE AICCSA*, 2008, pp. 941–942.
11. Y. Li, Z. Li, M. Chiang, and A. R. Calderbank, "Energy-efficient video transmission scheduling for wireless peer-to-peer live streaming," in *Proc. IEEE CCNC*, 2009, pp. 1–5.
12. M. Ismail, W. Zhuang, and S. Elhedhli, "Energy and content aware multi-homing video transmission in heterogeneous networks," *IEEE Transactions on Wireless Communications*, vol. 12, no. 7, pp. 3600–3610, July 2013.
13. Q. Yan, M. Li, Z. Yang, W. Lou, and H. Zhai, "Throughput analysis of cooperative mobile content distribution in vehicular network using symbol level network coding," *IEEE Journal on Selected Areas in Communications*, vol. 30, no. 2, pp. 484–492, Feb. 2012.
14. L. Zhu, F. Yu, B. Ning, and T. Tang, "Cross-layer design for video transmissions in metro passenger information systems," *IEEE Transactions on Vehicular Technology*, vol. 60, no. 3, pp. 1171–1181, March 2011.
15. H. Liang, L. X. Cai, D. Huang, X. Shen, and D. Peng, "An SMDP-based service model for interdomain resource allocation in mobile cloud networks," *IEEE Transactions on Vehicular Technology*, vol. 61, no. 5, pp. 2222–2232, June 2012.
16. N. Cheng, N. Lu, N. Zhang, J. Mark, and X. Shen, "Vehicle-assisted data delivery for smart grid: an optimal stopping approach," in *Proc. IEEE ICC*, June 2013, pp. 6184–6188.
17. N. Lu, N. Zhang, N. Cheng, X. Shen, W. Mark, and F. Bai "Vehicles meet infrastructure: Towards capacity-cost tradeoffs for vehicular access networks," *IEEE Transactions on Intelligent Transportation Systems*, vol. 14, no. 3, pp. 1266–1277, Sept. 2013.

18. H. Liang and W. Zhuang, "Efficient on-demand data service delivery to high-speed trains in cellular/infostation integrated networks," *IEEE Journal on Selected Areas in Communications*, vol. 30, no. 4, pp. 780–791, May 2012.
19. Y. Alayev, F. Chen, Y. Hou, M. P. Johnson, A. Bar-Noy, T. La Porta, and K. K. Leung, "Throughput maximization in mobile WSN scheduling with power control and rate selection," in *Proc. IEEE DCOSS*, 2012, pp. 33–40.
20. N. Zhang, N. Lu, N. Cheng, J. W. Mark, and X. Shen, "Cooperative spectrum access towards secure information transfer for CRNs," *IEEE Journal on Selected Areas in Communications*, vol. 31, no. 11, pp. 2453–2464, Nov. 2013.
21. N. Zhang, N. Cheng, N. Lu, H. Zhou, J. W. Mark, and X. Shen, "Risk-aware cooperative spectrum access for multi-channel cognitive radio networks," in *IEEE Journal on Selected Areas in Communications*, vol. 31, no. 11, Nov. 2013, pp. 2453–2464, vol. 32, no. 3, pp. 516–527, March 2013.
22. Y. Lee and H. D. Sherali, "Unrelated machine scheduling with time-window and machine downtime constraints: An application to a naval battle-group problem," *Annals of Operations Research*, vol. 50, no. 1, pp. 339–365, Dec. 1994.
23. F. Chen, M. P. Johnson, Y. Alayev, A. Bar-Noy, and T. F. La Porta, "Who, when, where: timeslot assignment to mobile clients," *IEEE Transactions on Mobile Computing*, vol. 11, no. 1, pp. 73–85, Jan. 2012.
24. G. Wifi, "Solar-powered internet in lascahobas, haiti," http://www.green-wifi.org/, 2011.
25. P. Kim and U. Geva, "Wind-powered wireless mesh network," http://ldt.stanford.edu/ educ 39109/POMI/MNet, 2011.
26. K. Baoqiang, C. Li, Z. Hongsong, and X. Yongjun, "Accurate energy model for WSN node and its optimal design," *Journal of Systems Engineering and Electronics*, vol. 19, no. 3, pp. 427–433, June 2008.
27. C. Ma and Y. Yang, "A battery-aware scheme for routing in wireless ad hoc networks," *Vehicular Technology, IEEE Transactions on*, vol. 60, no. 8, pp. 3919–3932, Oct. 2011.
28. J. Vazifehdan, R. Prasad, M. Jacobsson, and I. Niemegeers, "An analytical energy consumption model for packet transfer over wireless links," *IEEE Communications Letters*, vol. 16, pp. 30–33, Jan. 2012.
29. K. Ramachandran and B. Sikdar, "A population based approach to model the lifetime and energy distribution in battery constrained wireless sensor networks," *IEEE Journal on Selected Areas in Communications*, vol. 28, no. 4, pp. 576–586, Apr. 2010.
30. T. J. Kazmierski, G. V. Merrett, L. Wang, B. M. Al-Hashimi, A. S. Weddell, and I. N. Ayala-Garcia, "Modeling of wireless sensor nodes powered by tunable energy harvesters: Hdl-based approach," *Sensors Journal, IEEE*, vol. 12, no. 8, pp. 2680–2689, June 2012.
31. P. A. Chou and Z. Miao, "Rate-distortion optimized streaming of packetized media," *IEEE Transactions on Multimedia*, vol. 8, no. 2, pp. 390–404, Apr. 2006.
32. J. Chakareski and P. Frossard, "Rate-distortion optimized distributed packet scheduling of multiple video streams over shared communication resources," *IEEE Transactions on Multimedia*, vol. 8, no. 2, pp. 207–218, Apr. 2006.
33. T. H. Luan, L. X. Cai, and X. Shen, "Impact of network dynamics on user's video quality: Analytical framework and qos provision," *IEEE Transactions on Multimedia*, vol. 12, no. 1, pp. 64–78, Jan. 2010.
34. A. Sayegh, T. Todd, and M. Smadi, "Resource allocation and cost in hybrid solar/wind powered WLAN mesh nodes," in *Wireless Mesh Networks*, 2007, pp. 167–189.
35. N. Banerjee, M. D. Corner, and B. N. Levine, "An energy-efficient architecture for DTN throwboxes," in *Proc. IEEE INFOCOM*, 2007, pp. 776–784.
36. L. X. Cai, H. Poor, Y. Liu, T. H. Luan, X. Shen, and J. W. Mark, "Dimensioning network deployment and resource management in green mesh networks," *IEEE Wireless Communications*, vol. 18, no. 5, pp. 58–65, Oct. 2011.
37. Z. Zheng, L. Cai, R. Zhang, and X. Shen, "RNP-SA: Joint relay placement and sub-carrier allocation in wireless communication networks with sustainable energy," *IEEE Transactions on Wireless Communications*, vol. 11, no. 10, pp. 3818–3828, Oct. 2012.
38. Z. Zheng, L. X. Cai, and X. S. Shen, *Sustainable Wireless Networks*. SpringerBrief, 2013.

Chapter 3
Video Transmission Scheduling in Maritime Wideband Communication Networks

In this chapter, we develop a benchmark network model for vessel surveillance video uploading via a maritime wideband communication network. A broadband wireless network utilizing Time Division Multiple Access (TDMA) based media access control (MAC) protocol is exploited to provide a shore-side network framework, while the store-carry-and-forward routing scheme is adopted to mitigate the intermittent network connectivity in maritime communication scenarios. We are concerned about transmitting more important video packets to the administrative authority (i.e., TMP), based on the interpretation of throughput as the summation of weights of delivered video packets. Two offline scheduling algorithms are proposed, based on time-capacity mapping approach to convert the original time-based non-continuous resource allocation problem to a capacity-based continuous problem. As a retreat when optimal algorithm is out of reach, it is mathematically proved that the IGTJRS algorithm has an approximation ratio of 2, and a time complexity of $O(n^2)$. Simulation results validate the proposed algorithms, by obtaining the real ship route traces from navigation software BLM-Shipping.

3.1 System Model

We take into account of the scenario that a vessel creates surveillance videos periodically, sailing from origin port to destination. Video clips are divided into packets and uploaded to a content server of administrative agencies by infostations deployed along route line, or stored in a DTN throw-box (It is a small, stationary and inexpensive device equipped with wireless interfaces and storage, acting as a relay to create more connection opportunities [1]), and then be carried and forwarded by other vessels. As many symbols are adopted, some important notations and definitions are listed in Table 3.1.

T. Yang, X. (Sherman) Shen, *Maritime Wideband Communication Networks,*   15
SpringerBriefs in Computer Science,
DOI 10.1007/978-3-319-07362-0_3, © The Author(s) 2014

Table 3.1 Notations and definitions

Symbol	Definition
$A_{h,k}$	The capacity (the maximum number of bytes that could be delivered) of the kth frame of the hth infostation.
c_n	The marker point of the virtual period
H	The number of shoreside infostations
K_h	The number of frames within the hth infostation
S_i	The job instances of job i
t	The upper bound of all deadlines
T_F	Frame duration
$T_I(T_O)$	The starting (ending) time of the vessel
$T_h^i(T_h^o)$	The time for a vessel get into (come out of) the coverage of the hth infostation
$r_{jm}(d_{jm})(p_{jm})(b_{jm})(e_{jm})$	The release time (deadline) (processing time) (beginning time) (ending time) of video packet j proceeded on vessel m
$r_j^c(d_j^c)(p_j^c)$	The capacity of the release time (deadline) (processing time)
x_{jb_j}	Job j executed at the job interval $(b_j, b_j + p_j)$ indicator
w_j	Weight of job j

3.1.1 Store-Carry-and-Forward Routing

The network topologies are shown in Fig. 3.1. The sailing period of the vessel is $[T_I, T_O]$. H infostations deployed shore-side intermittently are deputed as base stations, whereas a broadband wireless network utilizing TDMA MAC protocol provides seamless coverages within the communication range of infostations. Meanwhile, the vessels are viewed as subscriber stations. A wireline/wireless network could connect the infostations to content servers of administrative authorities. Vessels could only communicate with infostations during the available time windows. It is obvious that several vessels might pass through the same crossing point with DTN throw-box, presumably not at the same time. Therefore, vessels can be engaged in cooperative transmissions for more efficient packets delivery via the DTN throw-box. In this brief, we only consider cargo vessels and passenger vessels sailing on predetermined and fixed route lines, in such a way that the schedule of vessels are comparatively stable and known *a priori*. However, the scenario of vessels with lower tonnage (such as fishing boats) might relate to stochastic modeling and will be further studied in the future work.

3.1.2 Network Resources

The network resource is defined as frames, within which the video packets can be delivered. Therefore, we utilize the IEEE standard 802.16/TDMA MAC frame structure to provide packets delivery service between vessel and infostations. The duration in which a vessel is within the coverage of the hth infostation is divided into frames with equal duration T_F. T_h^i and T_h^o denote respectively the times for a vessel

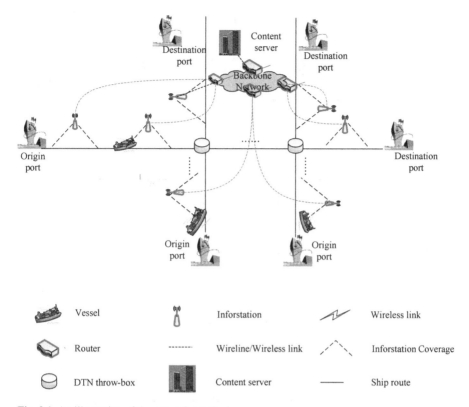

Fig. 3.1 An illustration of the network topologies

to get into and come out of the coverage of the hth ($h \in [1, \cdots , H]$) infostation. $A_{h,k}$ represents the maximum number of bytes could be transmitted (i.e., capacity) within the kth frame of the hth infostation. $T_h^i (T_h^o)$ means the time for a vessel get into (come out of) the coverage of the hth infostation. Figure 3.2 indicates network resources and the time-capacity mapping method.

3.1.3 Video Service

Surveillance video clips are separated into packets, and each packet has its release time, deadline, and weight. Weight reveals the importance to the video quality, as well as the significance to administrative authority. When the packet is transmitted before the deadline, the weight could be achieved. r_j, d_j, b_j, e_j, w_j and p_j represent the release time, deadline, beginning time, ending time, weight and processing time for job j. Denote $j \in \{1, \cdots , n\}$ as jobs, n as the number of jobs, $b_j \in \{1, \cdots , t\}$ as the beginning time of job j and $u \in \{1, \cdots , t\}$ as the beginning time of another job without scheduled overlapping with job j. Figure 3.3 shows video clips and packets.

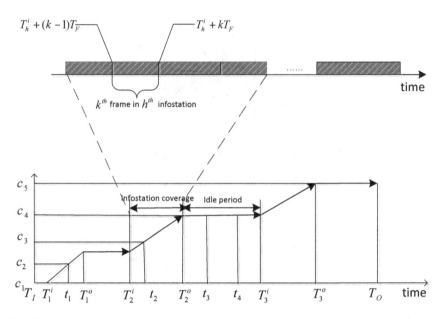

Fig. 3.2 Network resources and time-capacity mapping method

Fig. 3.3 Video clips and packets

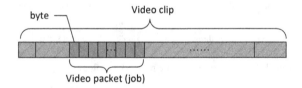

3.1.4   Time-Capacity Mapping

Time-capacity mapping method is adopted to transform the original time-based intermittent network connectivity into a virtually capacity-based continuous scenario [2]. The time indices are mapped into virtually cumulative capacity values, as shown in Fig. 3.2. The time-capacity mapping function $f(t):[T_I, T_o] \rightarrow [0, 1, \cdots \sum_{h=1}^{H} \sum_{k=1}^{K} A_{h,k}]$ is given by

$$f(t) = \begin{cases} \sum_{m=1}^{(t-T_{h_t}^i)/T_F} A_{h_t,m} + \sum_{l=1}^{h_t-1} \sum_{m=1}^{K_l} A_{l,m}, & \text{if } h_t \geq 1 \text{ and } T_{h_t}^i \leq t \leq T_{h_t}^o \\ \sum_{l=1}^{h_t} \sum_{m=1}^{K_l} A_{l,m}, & \text{otherwise} \end{cases}$$

$$(3.1)$$

where $h_t = \arg\max_h\{T_h^i \leq t\}$, if $T_{h_t}^i \leq t \leq T_{h_t}^o$, and $h_t = 0$ otherwise. Applying the time-capacity mapping method, the resource allocation issue could be converted from time based scheduling to capacity based scheduling over a continuous horizon [2], then the job-machine scheduling theory could be utilized to solve this problem with low complexity.

3.2 Problem Formulation

With regard to obtaining high video quality (i.e., maximizing the total weights of the accomplished packets), our scheme is scheduling video delivery with cooperation between vessels, revolving the deadlines and weights of the packets. And offline algorithms will be considered in this chapter. Job-machine scheduling model is utilized to schedule different video packets transmitting in different frames. With respect to job-machine scheduling method, vessels and video packets act as machines and jobs, respectively. Video packets transmitted by vessels corresponds to jobs performed on machines, i.e., jobs $\mathcal{J} = \{J_1, \cdots, J_n\}$ executed on machines $\mathcal{M} = \{M_1, \cdots, M_m\}$. We define family as a set of job instances execution possibility within release time and deadline, and no more than one job instance could be executed at the same time [3]. *job instances* are defined as a set of cases within a family, that could be executed during its release time and deadline. Let job instance S_i as a quadruple of the following variables: (family, value, beginning, ending), i.e., (i, w_i, b_i, e_i). Each job instance in one family has the same weight w_i. And at most one job instance of one family can be scheduled.

3.2.1 Formulation of Multi-Vessel Cooperation Delivery

In multi-vessel cooperation delivery scenario, when one vessel passing by the DTN throw-box node, the vessel could choose to forward the packets to the node. And then the DTN throw-box stores the data until other vessels pass by the node, they could carry and try to deliver packets with larger weights. Multi-vessel delivery, which corresponds to multi-machine scheduling, makes use of the different routes and infostations, and thus increase the chance of transmission.

Let r_{jm} and d_{jm} denote the release time and deadline for job j on machine m, where $j \in \mathcal{J}$ represents jobs and $m \in \mathcal{M}$ represents machines. Let w_{jm}, p_{jm}, and b_{jm} be the weight, processing time, and beginning time of job j on machine m, respectively. Apparently, we have $r_{jm} \leq b_{jm}$ and $b_{jm} + p_{jm} \leq d_{jm}$.

Denote $x_{jmb_{jm}}$ as the decision variable indicates that whether job j is executed on machine m at job interval $[b_{jm}, b_{jm} + p_{jm}]$, which is given by

$$x_{jmb_{jm}} = \begin{cases} 1, & \text{if job } j \text{ is performed at job interval } [b_{jm}, b_{jm} + p_{jm}] \\ 0, & \text{otherwise.} \end{cases}$$

Then, the multi-vessel VTMP is formulated as follows:

$$\max \sum_{j=1}^{n} \sum_{b_{jm}=r_{jm}}^{d_{jm}-p_{jm}} w_{jm} \cdot x_{jmb_{jm}} \tag{3.2}$$

$$s.t. \sum_{j=1}^{n} \sum_{b_{jm}=u-p_{jm}+1}^{u} x_{jmb_{jm}} \leq 1, \forall m, u \tag{3.3}$$

$$\sum_{b_{jm}=r_{jm}}^{d_{jm}-p_{jm}} x_{jmb_{jm}} \leq 1, \forall j \tag{3.4}$$

$$x_{jmb_{jm}} \in \{0, 1\}. \tag{3.5}$$

Algorithm 2: *Feedback Algorithm*
1: $B_1 \leftarrow A_1 : (t_{release}, t_{deadline})$ 2: $B' \leftarrow B_1 \cup B$ 3: Let B_2 be the set of packets which cannot be scheduled on vessel 2   4: $C \leftarrow B_1 \cap B_2$ 5: $A_2 \leftarrow C : (t_{release}, t_{deadline})$ 6: $A_1 \leftarrow A_1 \backslash A_2$

3.3.1.2 Normal Information Delivery Scenario

Next, normal information delivery scenario will also be studied, in which both vessel 1 and vessel 2 have video packets (but not emergency information) to upload. The primary distinction with normal information case is that vessel 2 also has its own videos to deliver. We add a feedback algorithm to analyze whether the packets relayed from vessel 1 (calculated by TMTP algorithm of two machines) compete the same period with the original packets on vessel 2. If there is contention, the following feedback algorithm (**Algorithm 2**) could be engaged to choose the packets should not be relayed to vessel 2. We define A as a set of all the overlapping packets that should not be delivered by vessel 1; A_1 as a set of packets in A that could be transmitted to vessel 2 obtained by Algorithm 2; Denote $t_{release}$ and $t_{deadline}$ as the release time and deadline of packets; Let B_1 as the set of packets in set A_1 that should be transmitted to vessel 2 by using Algorithm 2; B is the original packet set of vessel 2; B' is the set when B_1 merges into B, i.e., the packet set on vessel 2 accepting the packets of vessel 1; B_2 is the set of packets that could not be transmitted by vessel 2; C is the intersection of set B_1 and B_2. Here, the ending time is stipulated that $e' > t' + \delta, \delta > 0$, since vessel 1 could help vessel 2 to transmit in this scenario.

3.3.2 Interval Graph Theory Based Job Relay Selection Algorithm

Next, a more efficient interval graph theory based job relay selection (IGTJRS) algorithm (**Algorithm 3**) is introduced. Step 3–7 is to check jobs intersection. If there is no overlapping between job instances, the instances with the earliest beginning times could be scheduled by this algorithm, as indicated in step 12–13. However, the emphasis and difficulty is the case that packets with inevitable overlapping, i.e., they could not be utterly delivered by vessel 1. Step 8–10 is to choose two sets of job instances, in which one is scheduled by vessel 1 and the other is scheduled by vessel 2.

Borrowing the definition of maximum weight 2-independent set (MW2IS) in interval graph [4], the VTMP issue is mapped to maximum weight 2-independent set selection issue. Interval graph is an intersection graph of a multi-set of intervals on the real line. It has one vertex for each interval in the set, and an edge between

---

Algorithm 3: *Interval Graph Theory Based Job Relay Selection Algorithm*

1: Definition: the same as Algorithm 1, and $M = \emptyset$
2: Detection two jobs whether intersect with each other:
3: **for** $\forall J_i \in \mathcal{J}, \forall J_j \in J \backslash J_i$ **do**
4:     Let $b_i \leftarrow \{b_i^1, b_i^2, \cdots b_i^k\}$ express beginning time of job i intervals,
       $b_j \leftarrow \{b_j^1, b_j^2, \cdots b_j^k\}$ express beginning time of job j intervals
5:     'job instance $k_{\min}^i \leftarrow \min\{b_i^k\}$, $k_{\max}^j \leftarrow \max\{e_j^k\}$, $\alpha \leftarrow k_i^{b\min}$, $\beta \leftarrow k_j^{e\max}$
6:     **if** $e_i^\alpha > b_j^\beta$ **then**
7:         Job i and j mutually intersect, $J_j \leftarrow (j, w_j, b_{j\min}, e_{j\min})$,
           $J_i \leftarrow (i, w_i, b_{i\min}, e_{i\min})$
8:         draw relative interval graph $G(V, E)$, then use Algorithm 5 to obtain two
           maximum weight interval sets Q_u' and Q_v' to be delivered in vessel 1 and
           vessel 2
9:         $Q' \leftarrow Q_m'$, Q_m' has the latest ending time, and $m \in \{u, v\}$
10:        vessel 2 $\leftarrow$ DTN throw-box $\leftarrow$ the set Q'
11:    **else**
12:        Job i and j not mutually intersect, Find $(j, w_j, \widetilde{b}_j, \widetilde{e}_j), (i, w_i, \widetilde{b}_i, \widetilde{e}_i) \rightarrow M$
           with minimum $\widetilde{b}$
13:        $M \leftarrow M \cup (j, w_j, \widetilde{b}_j, \widetilde{e}_j) \cup (i, w_i, \widetilde{b}_i, \widetilde{e}_i)$
14:    **end if**
15: **end for**

---

every pair of vertices corresponds to intervals that intersect [5]. According to the intersection between job instances, an interval graph $G(V, E)$ is achieved.

In [4], the MW2IS algorithm could provide a collection of two sets which has maximum weights, without regard to the exact elements contained in each set. In order to adopt the scheduling issue in our scenario, we propose a modified MW2IS algorithm (**Algorithm 4**) to achieve the maximum two independent sets Q_u' and Q_v', as well as the exact elements in each set. *MWIS_IN_INTERVALS* is depicted in **Algorithm 5**. Finally, the one set which has the latest deadline in Q_u' and Q_v' will be chosen to be delivered to vessel 2. Regarding to multi-vessel scenario, it just needs to execute the above TMTP or IGTJRS algorithm multiple times, due to the offline characteristic.

3.3.3 Performance Analysis

We will analyze the performance of the proposed algorithms, in terms of the approximation ratio (i.e., the ratio of the throughput of the optimal schedule to that of the IGTJRS algorithm) and time complexity.

Algorithm 4: *Modified Maximum Weight 2-independent Set Algorithm*

1: **Input:** A set of weighted intervals $I \leftarrow \{i_1, i_2, \cdots i_n\}$ and the sorted endpoints list $L \leftarrow \{e_1, e_2, \cdots e_{2n}\}$.

2: **Output:** The MW2IS $Q_{\max 2}$ of I, two maximum weight interval sets Q'_u, Q'_v.

3: Step 1: $Q_{\max 2} \leftarrow \emptyset$; $Q_u \leftarrow \emptyset$; $Q_v \leftarrow \emptyset$; Set the initial value of $w(u,v), 0 \leq u, v \leq n$, to be 0.

4: Step 2: Compute each value of $w(u,v)$, $0 \leq u, v \leq n$, beginning from $w(0,v)$, $0 \leq v \leq n$, then $w(1,v)$, $0 \leq v \leq n, \cdots, w(n,v), 0 \leq v \leq n$, by algorithm MWIS_IN_INTERVALS and formula $w(u,v) = w(i_j) + \max\{\{w(u,x) \,|\, b_j > e_x \text{ and } u < x\} \cup \{w(x,v) \,|\, b_j > e_x \text{ and } x < v\}\}$, if $v > u$.

5: Step 3: Let $w(u_{n-1}, v_n) \leftarrow \psi(I)$; $Q_{\max 2} \leftarrow Q_{\max 2} \cup \{i_{u_{n-1}}, i_{v_n}\}$; $\psi(I) \leftarrow \psi(I) - (w(i_{u_{n-1}}) + w(i_{v_{n-1}}))$; $u_{n-1} \leftarrow u_1$; $v_n \leftarrow v_1$

6: **while** $\psi(I) > 0$ **do**

7:     Select a pair of intervals (i_{u2}, i_{v2}) with the constraints that $\max\{e_{u2}, e_{v2}\} < \max\{b_{u1}, b_{v1}\}$, $\min\{e_{u2}, e_{v2}\} < \min\{b_{u1}, b_{v1}\}$, and $w(u_2, v_2) = \psi(I)$.

8:     $Q_{\max 2} \leftarrow Q_{\max 2} \cup \{i_{u2}, i_{v2}\}$;

9:     $\psi(I) \leftarrow \psi(I) - (w(i_{u_2}) + w(i_{v_2}))$;

10:     $u_1 \leftarrow u_2$; $v_1 \leftarrow v_2$;

11:     $Q_u \leftarrow Q_u \cup \{i_{u1}\}$; $Q_v \leftarrow Q_v \cup \{i_{v1}\}$;

12: **end while**

13: Step 4: Let $Q_u \leftarrow \{Q_{u1}, Q_{u2}, \cdots Q_{um}\}$; $Q_v \leftarrow \{Q_{v1}, Q_{v2}, \cdots Q_{vm}\}$; $Q'_u \leftarrow \emptyset$; $Q'_v \leftarrow \emptyset$

14: $Q'_u \leftarrow Q'_u \cup \{Q_{u_1}\}$; $Q'_v \leftarrow Q'_v \cup \{Q_{v_1}\}$

15: **for** $\leftarrow 1$ *to* m **do**

16:     **if** $b_{Q_{u_i}} > e_{Q'_{u_{(i-1)}}}$ and $b_{Q_{v_i}} > e_{Q'_{v_{(i-1)}}}$ **then**

17:         $Q'_u \leftarrow Q'_u \cup \{Q_{u_i}\}$; $Q'_v \leftarrow Q'_v \cup \{Q_{v_i}\}$,

18:     **else**

19:         $Q'_v \leftarrow Q'_v \cup \{Q_{u_i}\}$; $Q'_u \leftarrow Q'_u \cup \{Q_{v_i}\}$

20:     **end if**

21: **end for**

Theorem 3.1 *The approximation ratio of the IGTJRS algorithm is 2.*

Due to the page limitation, we omit the details of the proof.

Regarding to time complexity, IGTJRS algorithm composes three steps of calculation:

Arrange Families: Array all the families $\mathcal{L}$ by the ending time. Binary search method will be adopted on $\mathcal{L}$ in $O(\log n)$ time. n denotes the number of jobs.

Check the Intersection: Linear search is used, with a time complexity $O(n)$.

Scheduling: Maximum weight 2-independent set algorithm solves the scheduling issue in $O(n^2)$ time.

Algorithm 5: *MWIS IN INTERVALS*

1: **Input:** A set of weighted intervals $I = \{i_1, i_2, \cdots i_n\}$ and the sorted endpoints list $L = \{e_1, e_2, \cdots e_{2n}\}$.
2: **Output:** The MWIS S_{max1} of I.
3: Step 1: $temp\_max \leftarrow 0$; $S_{max1} \leftarrow \emptyset$; $last\_interval \leftarrow 0$;
4: **for** $v \leftarrow 1$ to n **do**
5:     $\chi(v) \leftarrow 0$
6: **end for**
7: Step 2:
8: **for** $i \leftarrow 1$ to $2n$ **do**
9:     **if** e_i is a left endpoint of interval i_c **then**
10:        $\chi(c) \leftarrow temp\_max$
11:        **if** e_i is a right endpoint of interval i_c **then**
12:            **if** $\chi(c) > temp\_max$ **then**
13:                $temp\_max \leftarrow \chi(c)$;
14:                $last\_interval \leftarrow c$;
15:            **end if**
16:        **end if**
17:    **end if**
18: **end for**
19: Step 3: $S_{max1} \leftarrow S_{max1} \cup \{i_{last\_interval}\}$;
    $temp\_max \leftarrow temp\_max - w\{i_{last\_interval}\}$
20: **for** $v \leftarrow last\_interval - 1$ to 1 **do**
21:     **if** $\chi(v) = temp\_max$ and $e_v < a_{last\_interval}$ **then**
22:        $S_{max1} \leftarrow S_{max1} \cup \{i_v\}$;
23:        $temp\_max \leftarrow temp\_max - w\{i_v\}$;
24:        $last\_interval \leftarrow v$;
25:    **end if**
26: **end for**

Hence, the IGTJRS algorithm runs in $O(n^2)$ time, while TMTP algorithm runs in $O(2tn \log \log (2t))$ time. Here, t indicates the latest job deadline with $t \gg n$, since one job generally occupies multiple timeslots. Therefore, the IGTJRS algorithm has better performance in time complexity than TMTP algorithm.

3.4 Performance Evaluation

We will verify the proposed algorithms, based on the real traces of vessels in the area of Singapore Harbor which are captured from Navigation software BLM-Shipping [6]. Table 3.2 lists the time-position traces of vessels "Rainbow1" and "Secret". Figure 3.4a and Fig. 3.4b indicate the traces of vessels "Rainbow1" and "Secret". The infostations are evenly arranged along the shore-side. The DTN throw-boxes

Table 3.2 The time-position traces of vessels

Time(1)	"Rainbow1" position(1)	Time(2)	"Secret" position(2)	⋯
⋯	⋯	⋯	⋯	⋯
19 : 56	$1°13'54''N\,103°53'32''E$	19 : 37	$1°05'45''N\,103°44'01''E$	⋯
20 : 05	$1°12'20''N\,103°52'16''E$	19 : 52	$1°08'23''N\,103°46'20''E$	⋯
20 : 28	$1°10'31''N\,103°47'57''E$	19 : 58	$1°08'51''N\,103°46'46''E$	⋯
20 : 40	$1°09'10''N\,103°45'34''E$	20 : 03	$1°09'31''N\,103°47'38''E$	⋯
20 : 46	$1°08'51''N\,103°44'27''E$	20 : 19	$1°10'26''N\,103°49'35''E$	⋯
⋯	⋯	⋯	⋯	⋯
Time(3)	"Gloden Rise" position(3)	Time(4)	"Ayer" position(4)	⋯
⋯	⋯	⋯	⋯	⋯
19 : 58	$1°10'30''N\,103°52'20''E$	20 : 16	$1°07'46''N\,103°47'05''E$	⋯
20 : 06	$1°11'43''N\,103°51'52''E$	20 : 31	$1°08'21''N\,103°46'21''E$	⋯
20 : 11	$1°12'20''N\,103°51'19''E$	20 : 42	$1°09'57''N\,103°46'08''E$	⋯
20 : 18	$1°12'54''N\,103°49'51''E$	20 : 48	$1°10'12''N\,103°45'21''E$	⋯
20 : 25	$1°13'12''N\,103°48'15''E$	20 : 56	$1°10'32''N\,103°44'53''E$	⋯
20 : 25	$1°13'12''N\,103°48'15''E$	20 : 56	$1°10'32''N\,103°44'53''E$	⋯
⋯	⋯	⋯	⋯	⋯

are placed at the crossing points of vessels routes which is known *a priori*. And the simulation parameters are listed in Table 3.3.

We could only obtain the discrete points of traces from the software, therefore synthetic vessel trace method should be used in the simulations. Curve fitting is implemented with the hypothesis that there is always straight line between two adjacent points. Since the earth is round, the following formulas should be used to calculate the great circle distance S according to navigation science. Denote (φ_1, θ_1), (φ_2, θ_2) are the locations of two vessels, with φ and θ as the latitude and longitude respectively. We have

$$\cos S = \sin \varphi_1 \cdot \sin \varphi_2 + \cos \varphi_1 \cdot \cos \varphi_2 \cdot \cos D\lambda \qquad (3.10)$$

$$D\lambda = \theta_2 - \theta_1. \qquad (3.11)$$

Then the great circle distance formulas $S = \arccos(\cos S)$ could be utilized [7].

The proposed algorithms are verified by comparison with some other existing maritime communication algorithms [8; 9] and some classic scheduling algorithms i.e., Deadline (job with the earliest deadline will be scheduled first), First-input-first-output (FIFO) (job with the earliest release time will be scheduled first), Weight (job with the largest weight is scheduled first), and Single (noncooperative between two-vessel). Normalized throughput is defined as the ratio of the throughput of successfully transmitted packets to the throughput of total packets. In order to compare with the method in [8], whereas vessels utilizing the mesh network rather than DTN to deliver data, we assume a vessel could address data only when it is within the coverage of an infostation, and apply Deadline scheduling in the simulation. The literature [9] utilizes DTN mechanism and delivers the stored packets if it is possible. Nonetheless, there is no DTN throw-box, thus the opportunity of successfully delivered may be less than our proposed framework.

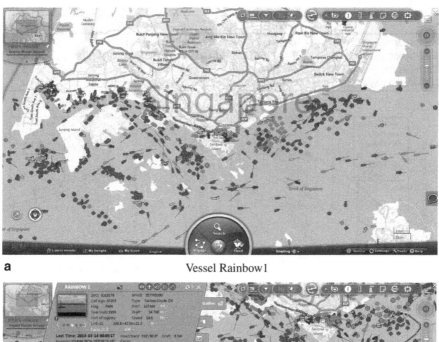

a   Vessel Rainbow1

b   Vessel Secret

Fig. 3.4 The traces of vessels Rainbow1 and Secret

The two-vessel scenario will be studied first, assuming that vessel "Rainbow1" has packets to deliver and vessel "Secret" undertakes a relay. In TMTP algorithm, we assume that vessel "Secret" has all the packets that vessel "Rainbow1" generates and delivers at the beginning of the simulation, which is not realistic. Thus, the performance of TMTP could be taken into account as the upper bound of all DTN algorithms. Single algorithm means noncooperative transmission that TMTP algorithm is utilized just for single machine. From Fig. 3.5, it is clear that cooperative schemes IGTJRS

Table 3.3 Simulation parameters

Name	Value	Name	Value
Packet size S_p	100 bytes	System bandwidth	10 MHz
Noise spectral density	174 dBm/Hz	Transmit antenna height h_{tx}	10 m
Video bitrate	0.47 Mbps	Frame duration T_F	5 ms
UE transmit power P_{tx}	23 dBm	Receive antenna height h_{rv}	50 m

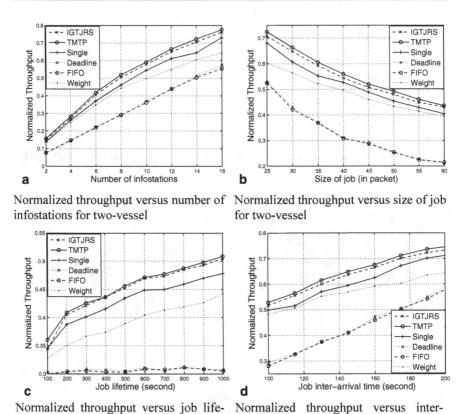

a Normalized throughput versus number of infostations for two-vessel

b Normalized throughput versus size of job for two-vessel

c Normalized throughput versus job lifetime for two-vessel

d Normalized throughput versus inter-arrival time for two-vessel

Fig. 3.5 Simulation results for two-vessel scenario

and TMTP outperform noncooperative Single scheme. Furthermore, IGTJRS algorithm obtains almost the same performance as TMTP algorithm, and obviously outperforms the other three cooperative algorithms (i.e., Deadline, FIFO, and Weight). Figure 3.5a indicates normalized throughput versus the number of infostations. The normalized throughput increases along with the increment of infostations number, since there are more chances to transmit. Figure 3.5b shows the impact of size of job. A larger size of job means the video packet requires more frame resource to transmit, which causes the total weights of transmitted packets decreasing. The normalized throughput versus job lifetime is indicated in Fig. 3.5c . Longer job lifetime brings

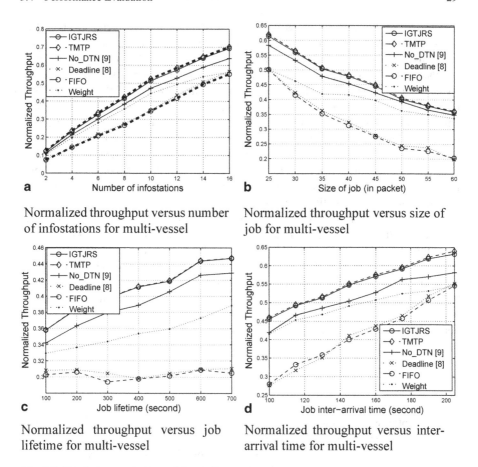

a Normalized throughput versus number of infostations for multi-vessel

b Normalized throughput versus size of job for multi-vessel

c Normalized throughput versus job lifetime for multi-vessel

d Normalized throughput versus inter-arrival time for multi-vessel

Fig. 3.6 Simulation results for multi-vessel scenario

in more non-overlapping job intervals and higher normalized throughput. From Fig. 3.5d, we could notice that the normalized throughput increases along with the job inter-arrival time increases, since the number of non-overlapping jobs increases.

Then, we will study the multi-vessel scenario in Fig. 3.6. In this scenario, vessels all have their own video packets to transmit, and may play as a DTN relay to help transmitting packets if it is available. The results will also be compared with those in [9] and [8]. Similar to two-vessel scenario, in TMTP algorithm, the assisting vessels are assumed to carry all the packets from original vessel in the beginning of the simulation. Hence, TMTP algorithm could still be considered to obtain the upper bound of performance. The algorithm in [8] does not perform as well as TMTP, IGTJRS, and [9] because it does not utilize DTN scheme. Although the algorithm proposed in [9] achieves better performance than [8], it performs worse than TMTP and IGTJRS since it does not apply the DTN throw-box. In Fig. 3.6a, normalized throughput versus the number of infostations is implemented. It could be seen that the normalized throughput increases as the infostations number increases. IGTJRS algorithm has almost the same performance as TMTP algorithm, and obviously outperforms the other four algorithms, which is identical with the above performance analysis. Figure 3.6b demonstrates normalized throughput versus the size of job (in packet). The number of scheduled jobs decreases, along with the job size increases. Figure 3.6c plots the normalized throughput versus job lifetime. With the job lifetime enlarges, the probability of non-overlapping job intervals increases and the performance of IGTJRS and TMTP algorithms is improved accordingly. While in Fig. 3.6d, the normalized throughput versus the job inter-arrival time is indicated, where the inter-arrival time varies from 100 to 200 s. It is seen that the normalized throughput increases as the increment of the arriving time, since the number of non-overlapping jobs also increases as the time span increases.

We could draw the following conclusions: (1) TMTP algorithm provides better performance than the other classic scheduling algorithms; (2) Compared with TMTP algorithm, IGTJRS algorithm could efficiently solve the scheduling issue with nearly the same performance; (3) Comparing with the existing maritime DTN/none-DTN algorithms, the cooperative IGTJRS and TMTP algorithms have better performance than noncooperative algorithms; (4) The normalized throughput increases as the number of infostations, job lifetime, and inter-arrival time increases, and (5) The normalized throughput decreases as the size of jobs increases.

3.5 Summary

In this chapter, we have shed light on the scheduling issue in maritime wireless communication network. Time-capacity mapping technique is introduced to transform the original intermittent network connectivity scenario into a virtually continuous scenario. Based on job-machine scheduling method, two offline scheduling algorithms, TMTP algorithm and IGTJRS algorithm have been developed, with the goal of maximizing the weighted throughput of delivered video packets. IGTJRS algorithm has a 2-approximation ratio, and runs in $O(n^2)$ time. The performance of our proposed algorithms are verified, through comparison with classic scheduling algorithms and existed works. Results in this chapter can be applied to predict the network performance and provide guidance on the design and implementations for maritime wideband communication networks.

References

1. W. Zhao, Y. Chen, M. Ammar, M. Corner, B. Levine, and E. Zegura, "Capacity enhancement using throwboxes in DTNs," in *Proc. IEEE MASS*, 2006, pp. 31–40.
2. H. Liang, and W. Zhuang, "Efficient on-demand data service delivery to high-speed trains in cellular/infostation integrated networks," *IEEE Journal on Selected Areas in Communications*, vol. 30, no. 4, pp. 780–791, May 2012.

3. P. Berman, and B. DasGupta, "Multi-phase algorithms for throughput maximization for real-time scheduling," *Journal of Combinatorial Optimization*, vol. 4, no. 3, pp. 307–323, Sept. 2000.

4. J. Y. Hsiao, C. Y. Tang, and R. S. Chang, "An efficient algorithm for finding a maximum weight 2-independent set on interval graphs," *Information Processing Letters*, vol. 43, no. 5, pp. 229–235, Oct. 1992.

5. D. R. Fulkerson and O. A. Gross, "Incidence matrices and interval graphs," *Pacific J. Math*, vol. 15, no. 3, pp. 835–855, Nov. 1965.

6. B. I. G. Ltd, "User manual," http://www.boloomo.com/shippingMain.

7. Y. Li, A. C. Landsburg, R. A. Barr, and S. Calisal, "Improving ship maneuverability standards as a means for increasing ship controllability and safety," in *Proc. MTS/IEEE OCEANS*, 2005, pp. 1972–1981.

8. J. Shankar Pathmasuntharam, P.-Y. Kong, M.-T. Zhou, Y. Ge, H. Wang, C.-W. Ang, W. Su, and H. Harada, "TRITON: High speed maritime mesh networks," in *Proc.* IEEE PIMRC, 2008, pp. 1–5.

9. P. Kolios and L. Lambrinos, "Optimising file delivery in a maritime environment through inter-vessel connectivity predictions," in *Proc. IEEE WiMob*, 2012, pp. 777–783.

Chapter 4
Green Energy and Content Aware Scheduling in Maritime Wideband Communication Networks

The advances of green wireless networks have brought an alternative energy for maritime wideband networks, which could significantly reduce the expense of maritime wideband networks construction and maintenance. In Chap. 3, the framework for vessel surveillance video uploading is developed. However, it has not considered the energy issues in the network scenario, i.e., both energy charging and discharging processes are not considered, and thus it is not applicable to the green-energy-powered maritime wireless networks. With the emerging green energy, such as solar, wind and hydro, etc, it is possible to cost-effectively construct green-energy-powered maritime wireless networks, which can render significant benefits in terms of both energy supplement and cost consumption. However, green energy is dynamic and sometimes even intermittent, which shifts the fundamental design criterion of maritime wideband communication networks from energy efficiency to energy sustainability. The video transmission scheduling issues considering network throughput and energy sustainability of green-energy-powered maritime wideband communication networks will be investigated in this chapter, by modeling the energy buffer of infostations and DTN throw box as a $G/G/1$ queue.

4.1 System Model

WiMAX/store-carry-and-forward interworking maritime wireless networks are also devised to overcome the restrictions of long-distance connection at sea and discontinuous infostations deployment. Similar to Chap. 3, the network model is shown in Fig. 4.1. Figure 4.1a expresses the single vessel scenario, while Fig. 4.1b expresses the two vessels scenario. OFDM-based WiMAX technology is considered in this chapter, which is more often adopted in wireless communication networks [1]. We still take uploading surveillance video clips from seagoing vessel to authority on land as an application scenario. Two paradigms of uplink will be taken into account, one is to deliver video packets via infostations deployed along the shore-side, the other is store-carry-and forward pattern borrowing DTN throw box. The infostations and the DTN throw boxes are powered by sustainable energy.

T. Yang, X. (Sherman) Shen, *Maritime Wideband Communication Networks,*   33
SpringerBriefs in Computer Science,
DOI 10.1007/978-3-319-07362-0_4, © The Author(s) 2014

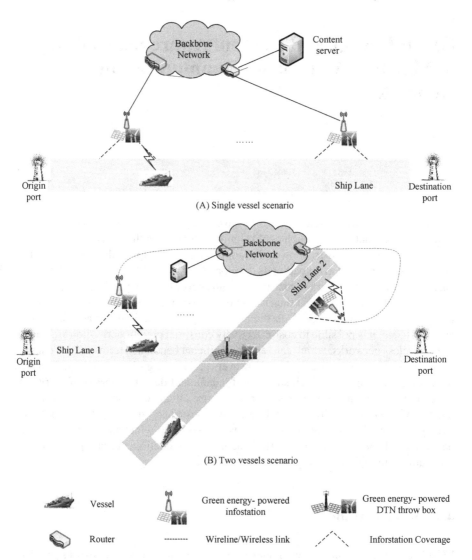

Fig. 4.1 An illustration of the network topologies

With regard to infostation and DTN throwbox, the harvested energy for packets transmission and energy complementarity is stored in a battery. Harvested energy will be charged to the energy buffer, while discharged by engaging video packets transmission. We model the energy buffer as a $G/G/1$ queue, i.e., the energy charging, the arrival and service time interval are independent and identically distributed ($i.i.d$) with the mean and variance of the inter-charging interval, noted as μ_a and υ_a, and the mean and variance of the energy inter-discharge interval are expressed as

Table 4.1 Notations and definitions

Symbol	Definition
$r_{jk}(d_{jk})(p_{jk})(s_{jk})$	The release time (deadline) (processing time) (starting time) of video packet j on vessel k
w_{jk}	Weight of packet j on vessel k
$x_{jks_{jk}}$	Binary variable denote whether packet j on vessel k is implemented at the time interval $[s_{jk}, s_{jk} + p_{jk}]$
$A(t)(L(t))$	The cumulative number of arriving and leaving energy
$X(t)$	A continuous process to approximate buffer size $R(t)$
$\alpha(\beta)$	Diffusion and drift diffusion coefficient
$\mu_a (\upsilon_a)$	The mean (variance) of energy inter-charging interval
$\mu_l (\upsilon_l)$	The mean (variance) of energy inter-discharge interval
x_0	The initial queue length (energy level)
$p(x, t; x_0)$	The conditional probability density function of the energy buffer size $X(t)$ at time t
$p_D(x, t; x_0)$	The probability density function of the buffer depletion duration D
$\mathcal{P}(0; x_0)$	The energy buffer depletion probability from x_0
$M_D(s)$	The moment generation function of D
$E(D) (Var(D))$	The mean(variance) of energy buffer depletion duration D
$\mathcal{P}(0; x_0)$	The energy buffer depletion probability from x_0
$F_D(T; x_0)$	The energy depletion probability before p_{jk} terminates

μ_l and υ_l, respectively. In this chapter, we focus on the scenario that the harvested energy may not be adequate to sustain all the traffic, that efficent video packets scheduling is significantly required. We summarize the main symbols used in this chapter in Table 4.1.

4.2   Energy and Content Aware Time-step-based Formulation

We are interested in video transmission scheduling targeting to maximize network throughput in green energy powered maritime wideband networks. Network throughput and the energy depletion probability should be conjointly considered as the metric of the formulated problem, based upon aforementioned analysis. Energy content aware video packets delivery schemes will be developed towards the maximum weights of delivered packets, namely energy and content aware vessel throughput maximization problem (EVTMP). Aiming to maximize the total weights of accomplished packets, meanwhile minimizing the probability that infostation and DTN throw box depleting energy when serves or stores traffic demands, the problem is formulated as follows:

Variable $x_{jks_{jk}}$ determines whether packet j is transmitted through vessel k at the time interval $[s_{jk}, s_{jk} + p_{jk}]$, i.e.,

$$x_{jks_{jk}} = \begin{cases} 1, & \text{if packet } j \text{ is performed on vessel } k \text{ at time} \\ & \text{interval } [s_{jk}, s_{jk} + p_{jk}] \\ 0, & \text{otherwise.} \end{cases}$$

Energy and content aware formulation is shown as

$$\max \sum_{j=1}^{n} \sum_{s_{jk}=r_{jk}}^{d_{jk}-p_{jk}} w_{jk} \cdot x_{jks_{jk}} \tag{4.1}$$

$$s.t. \sum_{j=1}^{n} \sum_{s_{jk}=u-p_{jk}+1}^{u} x_{jks_{jk}} \leq 1 \; k \in \{1,2\}, \forall u \tag{4.2}$$

$$\sum_{s_{jk}=r_{jk}}^{d_{jk}-p_{jk}} x_{jks_{jk}} \leq 1 \qquad\qquad k \in \{1,2\}, \forall j \tag{4.3}$$

$$x_{jks_{jk}} \in \{0,1\} \tag{4.4}$$

$$\mathcal{P}(0; x_0) < \varepsilon \qquad\qquad k \in \{1,2\}, \forall j \tag{4.5}$$

It involves four indices: packet, vessel, time step, and energy constraint (i.e., depletion probability), which is a $0-1$ integer nonlinear programming problem. The first constraint avoids multiple packets to be transmitted concurrently on the same vessel; the second constraint restricts that one packet could only be scheduled once; the third constraint indicates that the value of $x_{jks_{jk}}$ is 0 or 1; the fourth constraint shows the energy sustainability guarantee. $\mathcal{P}(0; x_0)$ is the energy depletion probability with initial energy x_0, which depicts the possibility that the infostations and DTN will deplete energy. $\varepsilon \ll 1$ is the threshold of energy depletion probability. Due to the page limitation, we omit the details of the proof that EVMTP considering energy restraint is $\mathcal{NP}$-complete.

4.3 Energy and Content Aware Video Transmission Scheduling Framework

We focus on video packets delivery scheduling to maximize the throughput of delivered video packets before the respective playback deadlines, subject to the limitation of energy. The framework takes into account of energy constraint, transient energy level, energy charging capability, and the depletion probability simultaneously, to meet the service requirements. We define a binary variable $x_{jks_{jk}}$, concerning about the video packet characteristics, opportunities to communicate with infostations, and the energy constraint. There is no efficient polynomial time solution, since the formulated problem is $\mathcal{NP}$-complete. We will design heuristic energy and content aware scheduling algorithms to maximize the weights of delivered packets with the energy sustainability constraint, satisfying the dynamics of the charging capability and video uploading requirements. Thus, two algorithms are proposed to conduct single vessel and two-vessel transmission scheduling respectively, i.e., an energy buffer-based decentralized online algorithm for single vessel and an energy buffer-based combinatorial decentralized-backward centralized algorithm for two vessels.

4.3.1 Leaky Bucket Energy Buffer Based Decentralized Online Algorithm for Single Vessel

We develop a decentralized algorithm to address the EVTMP problem. When video packet is created, a request message would be sent to the infostation by applying time slots to upload. The infostation will decide how to designate time slots to transmit the packet in terms of its knowledge, the initial energy level, and the energy charging capability of infostation. Then token will be allocated, or reject the uploading request.

Queueing Model of Energy Buffer The charging and discharging process model of green energy has been shown in [2]. Denote $A(t)$ and $L(t)$ as the cumulative number of charging and discharging energy unit at time t, respectively. $Q(0) = x_0$ is the initial energy level of infostation. The charged energy will be reserved in energy buffer, while the discharged energy will be consumed for video packet delivery. Thus the residual energy in the queue of energy buffer at time t is expressed as

$$Q(t) = A(t) - L(t). \tag{4.6}$$

It is critical to study the energy depletion duration of infostations, i.e., the duration from the initial level to depletion, which could be utilized to derive the probability of the infostations to run up energy after packet is uploaded. $G/G/1$ queue is applied to model the energy buffer, whereas random process is used to express energy charging and discharging model. Due to the dynamic characteristic of charging and discharging process, the infostation and DTN throw box might deplete energy when $Q(t) = 0$.

Borrowing from diffusion approximation theory [3, 4], we approximate the discrete buffer size $R(t)$ as a continuous process $X(t)$. Wiener-Levy process (or Brownian motion) model is resorted [5] as follows

$$dX(t) = X(t + dt) - X(t) = \beta dt + Z\sqrt{\alpha dt} \tag{4.7}$$

$Z \sim N(0, 1)$ denotes white Gaussian process with zero mean and unit variance. α and β denote drift and diffusion coefficients, expressed as follows

$$\begin{cases} \beta = E(\lim_{\Delta t \to 0} \frac{X(t)}{\Delta t}) = 1/\mu_a - 1/\mu_l \\ \alpha = Var(\lim_{\Delta t \to 0} \frac{X(t)}{\Delta t}) = v_a/\mu_a^3 + v_l/\mu_l^3. \end{cases} \tag{4.8}$$

The conditional probability density function (p.d.f) of the energy buffer size $X(t)$ at time t is expressed as

$$p(x, t; x_0) = \Pr(x \le X(t) \le x + dx \,|\, X(0) = x_0). \tag{4.9}$$

Resorting to Kolmogorov diffusion equation [5], it is given that

$$\frac{\partial p(x, t; x_0)}{\partial t} = \frac{\alpha}{2} \frac{\partial^2 p(x, t; x_0)}{\partial x^2} - \beta \frac{\partial p(x, t; x_0)}{\partial x}. \tag{4.10}$$

Since the queue length might not be negative, the queue length is derived as follows

$$p(x, 0; x_0) = \delta(x - x_0), \qquad t = 0 \tag{4.11}$$

$$p(0, t; x_0) = 0, \qquad t > 0, \tag{4.12}$$

where $\delta(x)$ denotes the Dirac delta function. By utilizing the images method [6, 7], the p.d.f of the energy buffer size is shown as

$$p(x, t; x_0) = \frac{\partial}{\partial x} \left\{ \Phi \left(\frac{x - x_0 - \beta t}{\sqrt{\alpha t}} \right) \right. \tag{4.13}$$
$$\left. - \exp \left(\frac{2\beta x}{\alpha} \right) \Phi \left(-\frac{x + x_0 + \beta t}{\sqrt{\alpha t}} \right) \right\},$$

where $\Phi(x)$ expressed as the standard normal integral, which is formulated as

$$\Phi(x) = \frac{1}{\sqrt{2\pi}} \int_{-\infty}^{x} \exp \left(-\frac{1}{2} y^2 \right) dy. \tag{4.14}$$

Denote $D(x_0) = \min\{t \geq 0 \,|\, X(0) = x_0, X(t) = 0\}$ as the energy buffer depletion duration with initial energy level x_0, the maximum energy duration for packet delivery is obtained. Thus, the probability density function of D could be retrieved resorting to the diffusion equation,

$$\frac{\partial p_D(x, t; x_0)}{\partial t} = \frac{\alpha}{2} \frac{\partial^2 p_D(x, t; x_0)}{\partial t^2} - \beta \frac{\partial p_D(x, t; x_0)}{\partial t}. \tag{4.15}$$

The conditional probability density function of the energy buffer depletion duration is

$$p_D(x, t; x_0) = \frac{x_0}{\sqrt{2\pi \alpha_D t^3}} \exp \left\{ -\frac{(x_0 + \beta_D t)^2}{2\alpha_D t} \right\}. \tag{4.16}$$

With Laplace transformation, the moment generation function of D is expressed as

$$M_D(s) = \exp \left\{ -\frac{x_0 \left(\beta_D + \sqrt{\beta_D^2 + 2\alpha_D s} \right)}{\alpha_D} \right\}. \tag{4.17}$$

The mean and variance of the energy buffer depletion duration D are expressed as

$$E(D) = -\frac{d}{ds} M_D(s) |_{s=0} = -\frac{x_0}{\beta_D} e^{-\frac{2\beta_D x_0}{\alpha_D}} \tag{4.18}$$

$$Var(D) = -\frac{d^2}{ds^2} M_D(s) |_{s=0} - E^2(D)$$
$$= e^{-\frac{2\beta_D x_0}{\alpha_D}} \left[2x_0 \beta_D^{-3} - x_0^2 \beta_D^{-2} \left(1 + e^{-\frac{2\beta_D x_0}{\alpha_D}} \right) \right]. \tag{4.19}$$

$\mathcal{P}(0; x_0)$ denotes the energy buffer depletion probability with initial level x_0

$$\mathcal{P}(0; x_0) = \lim_{D \to 0} \int_0^D p_D(x, t; x_0)dt = \lim_{s \to 0} M_D(s) \tag{4.20}$$

$$\mathcal{P}(0; x_0) = \begin{cases} 1, & for \ \beta_D < 0 \\ \exp\left\{-\frac{2x_0\beta_D}{\alpha_D}\right\}, & for \ \beta_D > 0 \end{cases} \tag{4.21}$$

It is observed from Eq. 4.21 that the energy buffer depletes with probability 1 when the energy charge rate is lower than or equal to the energy discharge rate $1/\mu_a \leq 1/\mu_l$. For the case that $1/\mu_a > 1/\mu_l$, the energy buffer depletion probability is determined by the initial energy level x_0 and the mean and variance of energy charging and discharging rates, etc.

The processing time of packet j on vessel k is denoted as p_{jk}, which means that the duration for uploading the video packet is p_{jk} time slots. $D(x_0)$ means the energy buffer depletion duration with initial energy x_0. The infostation will calculate the energy depletion probability before p_{jk} runs out,

$$F_D(T; x_0) = Pr(D \leq T) = \int_0^T p_D(x, t; x_0)dt. \tag{4.22}$$

Leaky Bucket Energy Buffer-Based Decentralized algorithm for Single Vessel
Algorithm 6 shows the Leaky bucket energy buffer-based decentralized algorithm for the single vessel delivery scenario. Each video packet is delivered with a token until the buffer is empty. Figure 4.2 shows the leaky bucket energy buffer diagram. The process of video packet generating and requesting could be modeled as Poisson distribution with λt, where λ denotes the average number of video packet arrivals per unit time. If a video packet arrives at time t, the following video packet arrives at time $t + \tau$, whereas τ means a random variable subjects to exponential distribution with parameter λ [8].

In decentralized algorithm, time slot reservation is allowed although the reservation cannot be guaranteed until the packet starts to be delivered. Even though the packet is in process, it could be also be canceled along with new packets arrival. The survival time T for the infostation is calculated, which means the depletion duration from its with initial energy level x_0.

$$F_D(T; x_0) = \int_0^T p_D(x, t; x_0)dt < \varepsilon \tag{4.23}$$

$$\int_0^T \left\{ -\frac{(x_0 + \beta t)^2}{2\alpha t} + \frac{1}{2}\left[\frac{(x_0 + \beta t)^2}{2\alpha t}\right]^2 \right\} \cdot \frac{x_0}{\sqrt{2\pi\alpha t^3}}dt \tag{4.24}$$

$$= \int_0^T \frac{x_0(x_0 + \beta t)^4}{8\alpha^4 t^{7/2}\sqrt{2\pi\alpha}} - \frac{x_0(x_0 + \beta t)^2}{2\alpha t^{5/2}\sqrt{2\pi\alpha}}dt \tag{4.25}$$

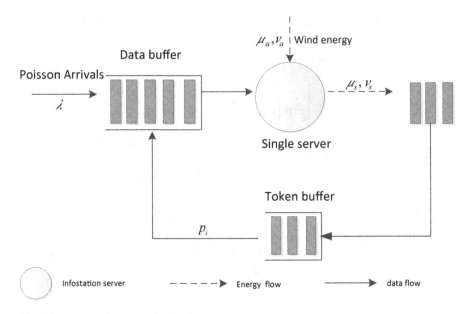

Fig. 4.2 Leaky bucket energy buffer diagram

$$= \frac{\beta^4 t^{3/2} x_0}{30\alpha^{5/2}} + \frac{2t^{1/2}\beta^3 x_0^2}{5\alpha^{5/2}} - \frac{2t^{1/2}\beta^2 x_0}{5\alpha^{3/2}} \bigg|_0^T \leq \varepsilon. \tag{4.26}$$

Since $p_D(x,t;x_0)$ is non-holonomic, the integral expression in Eq. 4.26 could not be achieved directly. Resorting to the first order of Taylor series expansion, the expression of T in Eq. 4.25 is approximated. Borrowing the solution of univariate cubic equation [9], the solution of T could be further obtained. Define a univariate cubic equation,

$$ax^3 + bx^2 + cx + d = 0, \quad a \neq 0, \tag{4.27}$$

then the solution of real number is:

$$x = -\frac{b}{3a} + \sqrt[3]{\frac{bc}{6a^2} - \frac{b^3}{27a^3} - \frac{d}{2a} + \sqrt{\left(\frac{bc}{6a^2} - \frac{b^3}{27a^3} - \frac{d}{2a}\right)^2 + \left(\frac{c}{3a} - \frac{b^2}{9a^2}\right)^3}}$$

$$+ \sqrt[3]{\frac{bc}{6a^2} - \frac{b^3}{27a^3} - \frac{d}{2a} - \sqrt{\left(\frac{bc}{6a^2} - \frac{b^3}{27a^3} - \frac{d}{2a}\right)^2 + \left(\frac{c}{3a} - \frac{b^2}{9a^2}\right)^3}}. \tag{4.28}$$

Once the infostation receives the request of video transmission from the vessel,

---

Algorithm 6: *Energy buffer based decentralized algorithm for single vessel*

---
1 $I \leftarrow \emptyset$;
2 *occupied* $\leftarrow 0$;
3 **for** *a new packet J_i arrives at time t* **do**
4   **if** *occupied* $+ p_j < T$ **then**
5   **if** $t >$ *occupied* **then**
6   schedule J_i at t;
7   *occupied* $\leftarrow t + p_i$;
8   **else**
9   there exist some scheduled packets J_r overlapped with J_i during l_r
10   **if** $w_i > \sum_r \left(1 + \frac{l_r}{p_r}\right) w_r$ **then**
11   replace packets J_r with J_i at t;
12                 $I \leftarrow I \cup \{J_r\}$;
13   *occupied* $\leftarrow t + p_i$;
14   **repeat**
15   reschedule $J_j \in I$ with highest $\frac{w_j}{p_j}$ at *occupied*;
16                     $I \leftarrow I \setminus \{J_j\}$;
17   *occupied* $\leftarrow$ *occupied* $+ p_j$;
18   **until** *no packets can be rescheduled*;
19   **else**
20   schedule J_i at *occupied*;
21   *occupied* $\leftarrow$ *occupied* $+ p_i$;
22   **end if**
23   **end if**
24   **end if**
25 **end for**

---

it would list all the packet instances that have not been transmitted. The algorithm considers four different cases:

case (a): When infostation receives a new request J_j, the packet is scheduled in descending order of $\frac{w_i}{p_i}$ until *occupied* $\geq t'$.

case (b): If that packet transmission is in progress, the packet J_j will be scheduled whenever the value of the current moment t adds the processing time p_j is no more than T.

case (c): In case that the value of current time t adds the processing time p_j is more than T, the packets that have already been scheduled will be preempted by packet J_j. The metric of preemption is $w_j > \sum w_r \cdot (1 + \frac{l_j}{p_r})$, i.e., the packet with higher weight preempts existing scheduled packets. Otherwise, it will be appended to the list of I.

The number of tokens available is indicated as i, and the token distribution rate is denoted by p_i. Actually, the processing time of each packet decides the token distribution rate. When a packet is delivered, the token will be reallocated for the next packet. The intervals of token distribution is ignored, so that the packets received tokens will be aligned in the data buffer. This algorithm guarantees no congestion in the single infostation.

4.3.2 Energy Buffer Based Combinatorial Decentralized-Backward Centralized Algorithm for Two Vessels

Cooperative transmission could further enhance the overall performance by creating more opportunities for packet uploading, utilizing the store-carry-and-forward mechanism, i.e., another vessel helping transmission via infostations en route. We assume two vessels are scheduled by a centralized server, which also schedules the green-energy-powered infostations and the DTN nodes. The server informs which packets could be stored in DTN throw box and be relayed by another vessel, by sending the acknowledgement or rejection messages. Given the hypothesis that there are always important video packets with higher weights on vessel 1 than vessel 2. Under the energy constraint, the key point in this case is how to allocate the uploading service between the two vessels, with the goal of maximizing the total weights of delivered video packets. In this case, we solve the packet delivery scheduling issue by taking into account of initial energy level, discharging and charging capacity. T decides whether the packet should be stored in DTN node or not, which is the processing time of all potential packets.

$$F_D(T; x_0) = \int_0^T p_D(t; x_0)dt < \varepsilon. \tag{4.29}$$

4.3.2.1 Maximum Carry Delay C

Then we will discuss the maximum carry delay C, assuming that the DTN throw box runs out energy after receiving the packets from vessel 1. The discharging process of the energy buffer is not mentioned in this scenario, i.e., $\mu_l = v_l = 0$. And the initial energy and charging parameters of the energy buffer are $x_o = 0$, μ_a and v_a. Thus, $\beta_C = 1/\mu_a$ and $\alpha_C = v_a/\mu_a^3$. Denote the minimal energy requirement of the DTN node as b. The maximum delay C passing data to vessel 2 is indicated as

$$C = \min \{t > 0 \,|X(0) = 0, X(t) = b\}. \tag{4.30}$$

Resorting to diffusion approximation, the probability density function of C is achieved as

$$\frac{\partial p_C(x, t; 0)}{\partial t} = \frac{\alpha}{2} \frac{\partial^2 p_C(x, t; 0)}{\partial t^2} - \beta \frac{\partial p_C(x, t; 0)}{\partial t} \tag{4.31}$$

$$p_C(x,0;0) = \delta(x), \quad t = 0 \tag{4.32}$$

$$p_C(b,t;0) = 0, \quad t > 0. \tag{4.33}$$

Utilizing the method of images [6, 7], the p.d.f expression is

$$p_C(x,t;0) = \frac{b}{\sqrt{2\pi\alpha_C t^3}} \exp\left\{-\frac{(b-\beta_C t)^2}{2\alpha_C t}\right\}. \tag{4.34}$$

And the moment generating function is indicated as

$$M_C(s) = \exp\left\{\frac{b}{\alpha_C}\left[\beta_C - \sqrt{\beta_C^2 + 2\alpha_C s}\right]\right\}. \tag{4.35}$$

Relatively, the mean and variance of the maximum carry delay C are expressed as

$$E(C) = -\frac{d}{ds}M_C(s)|_{s=0} = \frac{b}{\beta_C} = b\mu_a \tag{4.36}$$

$$Var(C) = -\frac{d^2}{ds^2}M_C(s)|_{s=0} - E^2(C) = bv_a. \tag{4.37}$$

Denote T_2 as the duration from the moment that DTN throw box receives packets to the moment that vessel 2 comes across it. The DTN throw box node will take into consideration of the probability that its energy reaches b before T_2.

$$F_C(T_2;0) = \Pr(C \le T_2) = \int_0^{T_2} p_C(x,t;0)dt \tag{4.38}$$

$$\int_0^{T_2}\left\{-\frac{(b-\beta t)^2}{2\alpha t} + \frac{1}{2}\left[\frac{(b-\beta t)^2}{2\alpha t}\right]^2\right\} \cdot \frac{b}{\sqrt{2\pi\alpha t^3}}dt \tag{4.39}$$

$$= \int_0^{T_2} \frac{b(b-\beta t)^4}{8\alpha^4 t^{7/2}\sqrt{2\pi\alpha}} - \frac{b(b-\beta t)^2}{2\alpha t^{5/2}\sqrt{2\pi\alpha}}dt \tag{4.40}$$

$$= \frac{\beta^4 t^{3/2}b}{30\alpha^{5/2}} - \frac{2t^{1/2}\beta^3 b^2}{5\alpha^{5/2}} - \frac{2t^{1/2}\beta^2 b}{5\alpha^{3/2}}\Big|_0^{T_2} \le \varepsilon. \tag{4.41}$$

As above mentioned, T_2 could be obtained by utilizing the truncating expansion equation of the Taylor series in Eq. 4.38, as well as the solution of univariate cubic equation in Eq. 4.40.

4.3.2.2 Energy Buffer Based Combinatorial Decentralized-Backward Centralized Algorithm

In this section, a combinatorial decentralized-backward centralized algorithm will be proposed. The algorithm is distributed before vessel 1 and vessel 2 arriving at the coverage of the DTN throw box. Then it becomes centralized algorithm when both vessels come upon within the communication range of DTN node, that DTN node schedule the traffic transmissions. Similar with Algorithm 6, the distributed algorithm is omitted here, and the centralized algorithm will be emphasized.

Denote $J = \{j_i\}, i \in [1, n]$ as the set of video packets that could not be transmitted by vessel 1; denote J_1 as the set of video packets stored in the DTN throw box, which is chosen by Algorithm 2; Denote x_0 as initial energy of the DTN throw box when vessel 1 delivers data to it; Let x_0' be the residual energy in the DTN throw box, and x_0'' be the energy level when vessel 2 communicates with the DTN throw box; Denote T_1 as the total length of video packets that will be delivered to the DTN throw box; Denote T_2 as the duration from the moment when vessel 1 delivers packets to DTN throw box to the moment when vessel 2 carries the packets; Denote T_3 as the duration for vessel 2 to receive the video packets. With the goal of maximizing energy and resource utility based on the green-energy modeling, the following scheduling scheme is exploited.

Step 1: Given the DTN initial energy x_0 and the residual energy x_0', T_1 is obtained from the following equation:

$$F_D(T_1; x_0) = \int_0^{T_1} p_D(x, t; x_0 - x_0')dt < \varepsilon. \qquad (4.42)$$

Step 2: If a packet's uploading time $T_J > T_1$, $\frac{w_i}{p_i}$ is utilized to sort the priority of packets which should be stored in the DTN node. Otherwise, $T_J < T_1$ shows that the energy is enough that all the packets could be stored in the DTN throw box.

Step 3: When vessel 2 carries data from the DTN throw box within T_2, the energy charging process is expressed as

$$F_C(T_2; 0) = \Pr(C \leq T_2) = \int_0^{T_2} p_C(x, t; 0)dt. \qquad (4.43)$$

We apply transformation $b \leftarrow x_0'' - x_0'$ to justify whether the energy is sufficient to implement the delivery, as well as whether the packets could be successfully delivered under energy constraint.

$$F_D(T_3; x_0) = \int_0^{T_3} p_D(x, t; x_0'')dt < \varepsilon. \qquad (4.44)$$

Utilizing $b \leftarrow x_0'' - x_0'$ and $x_0'' = x_0$, the value of x_0' could be obtained

$$F_C(T_2; 0) = \Pr(C \leq T_2) = \int_0^{T_2} p_C(x, t; 0)dt < \varepsilon. \qquad (4.45)$$

In Algorithm 7, vessel 2 assists vessel 1 to deliver unscheduled packets. Vessel 1 will determine which packets should be stored into DTN throw box at moment t. The number of tokens in the token buffer is i, and the token allocation rate is identical with the processing time p_i.

Algorithm 7: *Energy buffer based combinatorial decentralized-backward centralized algorithm for two vessels*

---

1 Two vessels decentralized algorithm is the same with Algorithm 1;
2 **Backward Centralized algorithm**

    Input: $x_0 = x_0'' = constant$; T_2; J is the set of total unscheduled packets in
               vessel 1, and T_J is total processing time of all the packets relatively; J_1
               is packets set to store in DTN node

    Output: x_0', J_1
3 $J_1 \leftarrow \emptyset$;
4 $occupied \leftarrow t$;
5 Calculate x_0' according to Eq. 4.40, Eq. 4.42 and Eq. 4.43; Calculate T_1
     according to Eq. 4.16, Eq. 4.25 and Eq. 4.42;
6 **for** *moment t vessel 1 store data into DTN node* **do**
7   **while** $J \neq \emptyset$ **do**
8   Schedule $J_i \in J$ which has the highest $\frac{w_i}{p_i}$;
9   **if** $occupied + p_i \leq T_1$ &$occupied + p_i \leq e_i$ **then**
10 $occupied \leftarrow occupied + p_i$;
11 $J \leftarrow J \setminus J_i \setminus \{ J_{e_i} < occupied + p_i \}$;
12 $J_1 \leftarrow J_1 \cup \{ J_i \}$;
13 $i \leftarrow i + 1$;
14   **else**
15 $J \leftarrow J \setminus \{ J_i \} \setminus \{ J_{e_i} < occupied + p_i \}$
16   **end if**
17   **end while**
18 **end for**

4.3.3   Time Complexity Analysis

Obviously, the worst case of packets overlapping decides the time complexity. For Algorithm 6, the worst case is that all the packets N are overlapped with packet 1. Thus the time complexity is indicated as $1 + 2 + \cdots + (N + 1) = \frac{N(N+1)}{2}$. Hence, the Algorithm 6 runs in $O(N^2)$ time, where N denotes as the maximum number of overlapping packets.

For Algorithm 7, it costs $O(\log N)$ time to engage the binary search method to schedule $J_i \in J$ with the highest $\frac{w_i}{p_i}$. N denotes the number of packets stored in DTN throw box.

Table 4.2 Time-position list of vessel rainbow1 and secret

Time(1)	Glory Hongkong position(1)	Time(2)	Chemroute Sun position(2)
. . .	. . .	. . .	. . .
16 : 49	$35°06'22''N129°26'33''E$	17 : 44	$35°09'26''N129°42'26''E$
17 : 10	$35°10'47''N129°29'15''E$	17 : 52	$35°11'10''N129°40'20''E$
17 : 18	$35°12'26''N129°29'15''E$	18 : 00	$35°12'23''N129°39'06''E$
17 : 29	$35°15'04''N129°31'29''E$	18 : 29	$35°17'26''N129°32'54''E$
17 : 57	$35°20'59''N129°34'50''E$	18 : 47	$35°19'51''N129°29'20''E$
. . .	. . .	. . .	. . .

4.4 Performance Evaluation

The performance of our proposed algorithms is evaluated based on the real vessel traces in the surrounding water of Busan Harbor of Korea. Vessel Glory Hongkong plays vessel 1, while vessel Chemroute Sun acts as vessel 2, being the relay. As above mentioned, the traces of vessel Glory Hongkong and Chemroute Sun could be obtained from BLM-Shipping navigation software. Table 4.2 shows the time-position information of the two vessels.

There are ten infostations deployed randomly along shoreline. The mean and variance of the energy charging interval are denoted as $\mu_a = 2.75$ and $\upsilon_a = 1.09$. Meanwhile, the mean and variance of the energy inter-discharging interval are denoted as $\mu_l = 4.35$ and $\upsilon_l = 11.1$. The proposed algorithms are evaluated by comparing with three classic scheduling algorithms, i.e., weight, deadline, and FIFO, with respect to normalized throughput versus deadline, arrival interval and processing time. All the algorithms take into consideration of survival time T, initial and consumption energy of infostations and DTN throw box.

Figure 4.3a indicates survival time T for infostation versus energy depletion probability $F_D(T; x_0)$. It shows that the longer the survival time, the higher the depletion probability. Figure 4.3b–d study the normalized throughput for single vessel scenario, in terms of packet deadline, inter-arrival time and processing time. In Fig. 4.3c, λ is utilized to show the characteristic of inter-arrival time since inter-arrival time τ follows exponential distribution with parameter λ. From the comparison, it is observed that our proposed algorithm significantly outperforms the other three algorithms. The proposed algorithm has the superiority to transmit the packets with maximum ratio of weight to processing time, i.e., $\frac{w_i}{p_i}$. That could guarantee that the packet with larger weight and less processing time could be scheduled first. For the other three algorithms, the weight algorithm outperforms deadline and FIFO algorithms, since this algorithm delivers the packets with the heaviest weight first. Contrastively, the deadline and FIFO algorithms only consider deadline and release time rather than weight, and thus results in the lower performance.

Figure 4.4a–c show normalized throughput versus deadline, inter-arrival time and processing time for two-vessel scenario. Our proposed algorithm significantly outperforms the other three algorithms. In Fig. 4.4a, more packets with higher weight could be delivered as the packet deadline is lengthened. It is observed from Fig. 4.4b

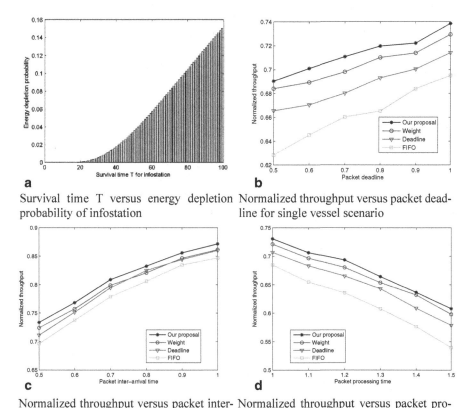

a Survival time T versus energy depletion probability of infostation

b Normalized throughput versus packet deadline for single vessel scenario

c Normalized throughput versus packet inter-arrival time for single vessel scenario

d Normalized throughput versus packet processing time for single vessel scenario

Fig. 4.3 Simulation results for single vessel scenario

that the normalized throughput is improved along with the inter-arrival time increasing, since the higher inter-arrival time brings in less number of overlapping packets and higher normalized throughput. Figure 4.4c shows that more processing time leads to lower normalized throughput, due to the limited time window to transmit.

Figures 4.5 and 4.6 indicate energy parameters versus normalized throughput for single and two-vessel scenarios, respectively. The proposed algorithms outperform the other three algorithms. From Figs. 4.5a and 4.6a, it is seen that the normalized throughput decreases as the increment of μ_a, since larger inter-charging interval causes insufficient energy for packet delivery. Figs. 4.5b and 4.6b indicate that the normalized throughput increases along with the increment of the mean of inter-discharging interval μ_l. Larger inter-discharging interval could bring in less energy consumption, which means that it is impossible for infostations to run out their energy.

Therefore, our proposed algorithms have better performance than the other classic algorithms, since our proposals take into account of both weight and processing time of the packets. The weight algorithm outperforms FIFO and deadline algorithms, since the factor weight has greater impact on the normalized throughput than deadline and release time.

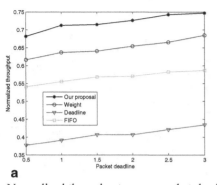

a

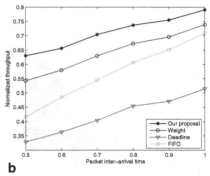

b

Normalized throughput versus packet dead-line for two-vessel scenario

Normalized throughput versus packet inter-arrival time for two-vessel scenario

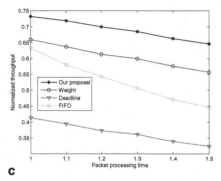

c

Normalized throughput versus packet pro-cessing time for two-vessel scenario

Fig. 4.4 Simulation results for two-vessel scenario

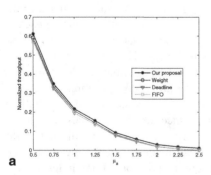

a

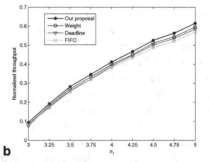

b

Normalized throughput versus the mean of energy inter-charging interval of infostatio for single vessel scenario

Normalized throughput versus the mean of energy inter-discharging interval of infostation for single vessel scenario

Fig. 4.5 Impact of energy parameters for single vessel scenario

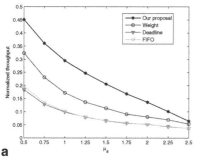
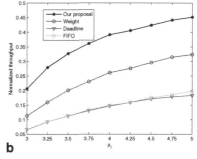

a Normalized throughput versus the mean of energy inter-charging interval of infostation for two-vessel scenario

b Normalized throughput versus the mean of energy inter-discharging interval of infostation for two-vessel scenario

Fig. 4.6 Impact of energy parameters for two-vessel scenario

4.5 Summary

In this chapter, the video transmission scheduling issue in terms of network throughput and energy sustainability in green-energy-powered maritime wideband networks is investigated. By modeling the green energy buffer as a $G/G/1$ queue, two algorithms have been developed for single vessel and two-vessel scenarios, with the goal of maximizing the network throughput, subject to the energy sustainability constraint. Exploiting real traces of vessels captured from BLM-Shipping software, extensive simulation results indicate that the proposed algorithms have better performance in both high network throughput and energy sustainability.

References

1. M. Mehrjoo, M. K. Awad, and X. S. Shen, "Resource allocation in OFDM-based WiMAX/in Book: WiMAX network planning and optimization," pp. 113–131, Apr. 2009.
2. L. X. Cai, Y. Liu, T. H. Luan, X. Shen, J. W. Mark, and H. V. Poor, "Adaptive resource management in sustainable energy powered wireless mesh networks," in *Proc. IEEE GLOBECOM*, 2011, pp. 1–5.
3. L. Kleinrock, *Queueing Systems: Volume 2: Computer Applications*, 1976.
4. A. O. Allen, *Probability, statistics, and queueing theory: with computer science applications*, 1990.
5. H. Kobayashi, "Application of the diffusion approximation to queueing networks i: Equilibrium queue distributions," *Journal of the ACM*, vol. 21, no. 2, pp. 316–328, Apr. 1974.
6. D. D. R. Cox and H. D. Miller, *The theory of stochastic processes*, 1977.
7. W. Feller, *An introduction to probability theory and its applications*, 2008.
8. F. A. Haight and F. A. Haight, "Handbook of the poisson distribution," 1967.
9. L. Hong, "A direct rigorous conversion from cartesian to geodetic coordinates based on the solution to univariate cubic equation," *Geomatics Spatial Information Technology*, vol. 6, pp. 48–50, June 2007.

Chapter 5
Conclusions and Future Directions

In this chapter, we summarize the main results presented in this monograph and highlight future research directions.

5.1 Conclusions

In this monograph, we have investigated the video transmission scheduling in maritime wideband communication networks. The main content of this brief is shown as follows:

- We have investigated the previous works related to maritime wideband communication networks and video transmission scheduling issues. Several research problems have been discussed including functions showcasing, potential applications, video transmission scheduling and energy modeling based video transmission scheduling under the premise of analyzing the unique characteristics of the dedicated maritime wideband communication networks.
- We have shed light on the video transmission scheduling schemes in maritime wideband communication networks. Time-capacity mapping technique is introduced to transform the original intermittent network connectivity scenario into a virtually continuous scenario. With the objective of maximizing the weighted throughput of the transmitted video packets, job-machine scheduling method is utilized to develop two offline scheduling algorithms. We show that the IGTJRS algorithm has a 2-approximation ratio, and runs in $O(n^2)$ time. Simulation results verify the proposed algorithms significantly outperforms the other classic and existing scheduling algorithms.
- We have addressed video transmission scheduling in terms of network throughput and energy sustainability in the green-energy-powered maritime wideband communication networks. Based on modeling green energy buffer as a $G/G/1$ queue, two algorithms are developed to solve the formulated energy and content aware vessel throughput maximization problem. Single vessel and two-vessel scenarios are both taken into account to maximize the network throughput, subject to the energy sustainability constraint. Exploiting real traces of vessels captured

T. Yang, X. (Sherman) Shen, *Maritime Wideband Communication Networks*, 51
SpringerBriefs in Computer Science,
DOI 10.1007/978-3-319-07362-0_5, © The Author(s) 2014

from BLM-Shipping software, extensive simulation results indicate that our algorithms have better performance in both high network throughput and energy sustainability.

5.2 Future Research Directions

We believe this is just the tentative beginning of maritime wideband communication networks study. The mobility model of vessel, dynamic energy harvest and heterogeneous traffic demand on the ocean lead to many new challenging research issues under the scenario of maritime wideband communication networks. We close this chapter with three additional thoughts on future research directions in this field.

- In this monograph, we consider passenger vessels and cargo vessels in an area with large navigation density. According to the Ships' Routing scheme recommended by IMO, most of the traces of passenger vessels and cargo vessels are comparatively deterministic or predictable, since the ship routes are relatively stable and known *a priori*. The analysis of vessels with lower tonnage (such as fishing boats) may involve in stochastic modeling and is left for our future work. Multi-vessel data transmission scheduling based on traffic flow analysis in different water area is a complicated issue in further research on maritime wideband networks. Thereby, routing scheme design under various network topologies should be further investigated.
- In this monograph, we suppose that the cooperative vessels have the responsibility to help the other vessels relaying data, jointly considering the total benefit. However, it is not realistic in practice. How to stimulate cooperation between the vessels in order to avoid selfishness is another interesting topic. Some advanced mathematical methods such as game theory will be utilized to develop stimulation mechanisms. In game theory, vessels are assumed to be selfish and rational, who are only interested in their own utilities [1]. By analyzing the game, a set of strategy can be determined such that each vessel can increase its utility by just changing its strategy. This issue makes the study closely related to game theory a demanding task.
- For maritime network performance, throughput is not the only metric. From the applications point of view, network delay and fairness are also important design aspects [2]. In the future work, we will refer to an interrelated throughput-delay-fairness (TDF) triplet. The transmission scheduling issue should be investigated in the novel maritime networks, considering the tradeoff between throughput, delay and fairness, however it will bring much more challenges.

References

1. T. Basar, G. J. Olsder, G. Clsder, T. Basar, T. Baser, and G. J. Olsder, *Dynamic noncooperative game theory*, 1995.
2. J.-G. Choi and S. Bahk, "Cell-throughput analysis of the proportional fair scheduler in the single-cell environment," *IEEE Transactions on Vehicular Technology*, vol. 56, no. 2, pp. 766–778, March 2007.